BEHAVIORAL
DEVELOPMENT
OF NONHUMAN PRIMATES

An Abstracted Bibliography

IFI DATA BASE LIBRARY

COMPUTER TECHNOLOGY
Logic, Memory, and Microprocessors — A Bibliography
A. H. Agajanian

ACOUSTIC EMISSION
A Bibliography with Abstracts
Thomas F. Drouillard

MICROELECTRONIC PACKAGING
A Bibliography
A. H. Agajanian

SPEECH COMMUNICATION AND THEATER ARTS
A Classified Bibliography of Theses and Dissertations, 1973—1978
Merilyn D. Merenda and James W. Polichak

ATOMIC GAS LASER TRANSITION DATA
A Critical Evaluation
William Ralph Bennett, Jr.

BEHAVIORAL DEVELOPMENT OF NONHUMAN PRIMATES
An Abstracted Bibliography
Faren R. Akins, Gillian S. Mace, John W. Hubbard, and Dianna L. Akins

MASTER TABLES FOR ELECTROMAGNETIC DEPTH SOUNDING
INTERPRETATION
Rajni K. Verma

BEHAVIORAL DEVELOPMENT OF NONHUMAN PRIMATES

An Abstracted Bibliography

Faren R. Akins
University of Santa Clara

Gillian S. Mace
San Jose State University

John W. Hubbard
University of Tennessee

and

Dianna L. Akins
Stanford University

IFI/PLENUM • NEW YORK-WASHINGTON-LONDON

Library of Congress Cataloging in Publication Data

Main entry under title:

Behavioral development of nonhuman primates: an abstracted bibliography.

(IFI data base library)
Includes index.
1. Primates — Behavior — Abstracts. 2. Primates — Behavior — Bibliography.
3. Mammals — Behavior — Abstracts. 4. Mammals — Behavior — Bibliography.
I. Akins, Faren R.
QL737.P9B385   156   79-26700
ISBN 0-306-65189-0

The preliminary phase of this research was supported by a
National Research Council Research Associateship to Faren R. Akins
while at the National Aeronautic and Space Administration, Ames
Research Center.

© 1980 IFI/Plenum Data Company
A Division of Plenum Publishing Corporation
227 West 17th Street, New York, N.Y. 10011

Printed in the United States of America

ACKNOWLEDGMENTS

The impetus for this volume was largely a result of the enthusiasm
and persistence of a fifteen member student research team organized
at NASA - Ames Research Center. The following individuals deserve
appropriate accolades for their excellent work in obtaining many
of the articles for this book. They are listed here along with
their educational affiliation at the time of their participation:

Christopher Atkinson   Dean Marsh
San Jose State University   Willow Glen High School

Theresa Freitas   Gloria Morales
University of Santa Clara   San Jose State University

John Gruver   Mike Myers
DeAnza Community College   West Valley Community College

John Hubbard   Raenel Neitz
University of Santa Clara   San Jose State University

Gisela Kahle   Bob Rathmell
San Jose State University   University of Santa Clara

Gillian Mace   Yvonne Rohan
San Jose State University   University of California,
                                    Santa Cruz

Kevin Madej
San Jose State University   Sharon Seyman
                                    Willow Glen High School

             Debbie Wessler
             DeAnza Community College

All of the people involved in this project wish the very best
to two other members of our team to whom this volume is dedicated:
Killer and Leroy.

PREFACE

The present volume represents the result of two years of work originally begun as a fifteen-member student project under my supervision at NASA - Ames Research Center, Moffett Field, California. As a means of acquainting team members with previous research related to our NASA experiments with long-term isolation and confinement effects upon nonhuman primate behavior a weekly meeting was arranged for students to orally present abstracts of various articles they had read. As the number of references increased we decided to expand our efforts through several computer searches of the psychological, biological, anthropological, and medical literature. Upon completion of our experiments at NASA, three of the team members and myself decided to take this basic foundation, update, expand and otherwise polish it into the present comprehensive reference tool we feel confident will be of value to investigators and scholars interested in the broad topic of nonhuman primate development as affected by early environmental influences. While ours is the only bibliography of this literature which includes both abstracts and indexing, several previous publications are worth noting as we found them particularly helpful in our own work. Those bibliographies, compiled by Agar and Mitchell (1973), Stoffer and Stoffer (1976), and Roy (1976, 1977), are excellent. In addition to the articles cited in these sources we have added approximately 400 more articles with abstracts and indexing. References included in this volume span a wide range of topics including the effects of social deprivation (both partial and total), dietary deficiencies, various types and degrees of stimulus deprivation, birth traumas, surgical procedures, and unusual rearing conditions. The abstracts which appear here are either those provided by the authors in their articles or were prepared by myself and my co-authors. There are four separate index sections to aid readers: species, topical area, author, and a listing of review or synthesis chapters and articles which pertain to certain broad aspects of this literature.

FAREN R. AKINS, Ph.D.

CONTENTS

1. Aakre, B., Strobel, D.A., Zimmermann, R.R., & Geist, C.R.
Reactions to intrinsic and extrinsic rewards in protein-malnourished
monkeys. Perceptual and Motor Skills, 1973, 36, 787-790.

> Ten rhesus monkeys were tested on a 12-part puzzle
> manipulation apparatus. Six of 10 Ss were maintained
> on a protein-deficient diet containing 3.5% casein by
> weight, while the remaining 4 Ss received a high-
> protein diet containing 25% casein by weight. The high-
> protein animals had an elevated manipulation rate as
> compared with low-protein fed Ss when the only source
> of reinforcement was the manipulation of the puzzles
> (intrinsic reward). However, when food (extrinsic
> reward) was introduced the low protein fed Ss manip-
> ulated at a level equal to or greater than the high-
> protein fed animals. Introduction of 100% and partial
> reinforcement conditions showed manipulation rates to
> be relatively consistent between the two groups.
> Extinction conditions, however, showed a significant
> difference between the high- and low-protein fed
> animals after partial reinforcement. With the removal
> of the food reward the low-protein fed animals exhibited
> a much more rapid reduction of manipulatory activity
> than the high-protein fed Ss. The more rapid extinction
> of the manipulation response by low-protein fed
> monkeys suggests that extrinsic reward is a much more
> salient variable for these Ss than intrinsic reward.

2. Agar, M., & Mitchell, G. Bibliography on deprivation and
separation, with special emphasis on primates. Abstracted in the
JSAS Catalog of Selected Documents in Psychology, 1973, 3, 79.
(M.S. No. 404)

> A 270 entry bibliography of studies concerned with
> separation and deprivation. Humans were the subjects
> in 135 of the studies, 111 of the studies dealt with
> prosimians, monkeys (primarily macaques), and apes
> (chimpanzees), and 15 of the studies involved other
> animals. Topics included are: Mother-infant separations
> and other forms of maternal deprivation, father-infant
> (or adult male-infant) separation, father-adolescent
> and mother-adolescent separation, peer separation,
> infant-preadolescent separation, adult male-female
> separation, separation of dominant male from group,
> and division and transplantation of members within
> a group. A special concern in many studies was
> separation brought on by death. Articles date from
> 1940 to 1973.

3.   Ainsworth, M.D.S. Discussion of papers by Suomi and Bowlby.
In   G. Serban and A. Kling (Eds.), Animal models in human
psychobiology. New York and London: Plenum Press, 1976, 37-47.

John Bowlby began his chapter (with acknowledgments to
Robert Hinde) with a statement of six ways in which
studies of animal behavior can contribute to a study
of human behavior. Two of these pertain especially
to experimental studies, two perhaps specifically to
naturalistic studies, and two to studies of both
types. This chapter first considers experimental studies
and their contributions with particular reference to
Dr. Suomi's chapter, and then turns to naturalistic
studies, referring both to Dr. Bowlby's paper and to
work of the author's. The pertinence to research into
human social development of naturalistic animal studies
is stressed; especially the contribution that an
ethological perspective has made to research with
humans.

4.   Akert, K., Orth, O.S., Harlow, H.F., & Schiltz, K.A.
Learned behavior of rhesus monkeys following neonatal bilateral
prefrontal lobotomy. Science, 1960, 132, 1944-1945.

Bilateral ablation of the dorsolateral frontal cortex in
the newborn rhesus failed to produce learning deficits
typically found in adolescent and adult animals with
bilateral lesions in the prefrontal areas. Tests
(delayed response, string tests, and discrimination
learning) were performed on the subjects 3 to 4 months
postoperatively, and no significant differences between
the controls and the animals with prefrontal lesions
were found at this time.

5.   Albert, S. & Rosenblum, L.A. The influence of gender and
rearing conditions on attachment and response to strangers.
Proceedings of the 5th International Congress of Primatology,
1974, 217-231.

In order to provide systematic data on the emergence
of selective, preferential response to mother and the
emergence of related or opposed reactions to strange
conspecifics, a series of tests were run on 15 bonnet
macaques (M. radiata) reared by their mother in group
settings and 4 infants of the same species raised

alone with their mothers. Although the group-reared
infants, both male and female, achieved reliable
preferential response to mother during the 2nd three
months of life, the females manifested consistently
higher preferences for mother as compared to stranger
when either was presented alone, or when both were
presented together. Furthermore, female infants,
within one to two months after emergence of very high
levels of preferential response to mother, showed a
relative wariness or avoidance of conspecific strangers.
In contrast to the group-reared infants, the infants
reared alone with their mothers failed to develop any
clear preferential response to mother as compared to
stranger. In addition, they failed at any age to
manifest any degree of wariness or relative avoidance
of conspecific strangers. No sex differences for
the infants reared alone with their mothers were found.

6. Alexander, B.K. The effects of early peer deprivation on the
juvenile behavior of rhesus monkeys (Doctoral dissertation,
University of Wisconsin, 1966). Dissertation Abstracts
International, 1967, 28(3).

This project was undertaken to test the hypothesis
that social contact with peers is a necessary prerequisite to normal social development in the rhesus
monkey (Macaca mulatta). The results suggest that
peer-deprivation treatment did not severely retard
the development of normal social behavior, and indicate
that the hypothesis that early contact with peers is
essential to all aspects of normal development in
the rhesus monkey must be modified.

7. Alexander, B.K., & Harlow, H.F. Social behavior of juvenile
rhesus monkeys subjected to different rearing conditions during
the first six months of life. Zoologische Jahrbucher
Physiologie, 1965, 71, 489-508.

Four groups of rhesus monkeys about a year and a half
old were tested on a battery of social behavior tests.
Three of the groups were allowed to interact with age
mates during the first six months of life but had
mothering conditions which ranged from no mothers,
to "motherless mothers", to normal mothers. The fourth
group was raised on surrogate mothers but had no chance

to interact with peers. All four groups were given
extensive social experience when they were between
6 and 18 months of age. Social testing was conducted
on the original groups, new groups formed by a dominance
criterion, the original groups reformed, and finally
on a second set of unfamiliar groups. The results
showed no consistent differences between the three
groups which were allowed to interact with peers in
infancy, but a generally inadequate development of
aggressive, affiliative, play, and sexual behavior
in the fourth group was found. These results support
data which indicate that peer experience early in life
is an essential prerequisite to normal social development in the rhesus monkey.

8.   Alexander, B.K. The effects of early peer-deprivation
on juvenile behavior of rhesus monkeys. (Doctoral Dissertation,
University of Wisconsin, 1966), Dissertation Abstracts International.

To test the hypothsis that social contact with peers is
a necessary prerequisite to normal social development
in the rhesus monkey, three four-member groups of infant
rhesus monkeys were studied. All three groups were
reared by their biological mothers and differed only
in the amount of peer contact permitted during the
first 8 months of life. The results of the tests
given at the end of the 8-month treatment phase indicated
that peer deprivation had not severely retarded the
development of normal social behavior. Although the
peer-deprived monkeys, in general, engaged in considerably more agressive behavior and less affiliative
behavior than did the controls, they were competent
at play, had adequate juvenile sexual patterns, and
appeared to have developed a normal group structure.

9.   Allen, J.R., McWey, P.J., & Suomi, S.J. Pathobiological and
behavioral effects of lead intoxication in the infant rhesus
monkey. Environmental Health Perspectives, 1974, 7, 239-246.

When infant rhesus monkeys were exposed to lead via
the addition of lead acetate (0.5-9mg/kg body weight)
to their formula or by the consumption of lead particles
from lead-based surrogate mothers, they developed
symptoms of lead intoxication within 6 weeks. Seizures,

muscular tremors, and altered social interaction were
the predominant changes. Visual impairment was also
apparent in the more severely affected animals. In
the animals showing obvious symptoms lead levels varied
between 300 to 500 µg/100 ml of blood. Even in those
animals having blood lead levels below 100 µg, hyperactivity and insomnia were observed. When the exposure
to lead was eliminated, seizures subsided and visual
impairment was reduced; however, the abnormal social
interaction persisted. These animals also experienced
a gradual decline in hematocrit and hemoglobin values
during the period of examination. Liver and kidney
biopsies obtained from these lead-exposed animals
revealed characteristic intranuclear inclusions. When
adolescent and adult monkeys were exposed to doses of
lead acetate similar to those employed in the infant
experiments, lead levels in excess of 200 µg/100 ml
of blood were recorded. However, there were no obvious
behavioral abnormalities observed. There were, however,
numerous lead inclusion bodies in kidney biopsy
specimens from these animals.

10. Allyn, G., Deyme, A. & Begue, I. Self-fighting syndrome in
Macaques: 1. A representative case study. Primates, 1976, 17(1),
1-22.

Spontaneous self-fighting behavior was studied clinically
in a Stumptail Macaque (Macaca arctoides) with a history
of prior social deprivation. Eighteen behavioral patterns
were studied and scored from vidio tapes which were
made of the subject under seven experimental conditions.
The results suggest that a general state of excitation
(present in this case as a result of external agression
against the animal being observed) was a triggering
factor in the self-fighting syndrome when the animal
could neither escape nor defend itself.

11. Altman, S.A. Sociobiology of rhesus monkeys: IV. Testing
Mason's hypothesis of sex differences in affective behavior.
Behavior, 1968, 32, 49-69.

To test the hypothesis of Mason, Green and Posepanko
(1960) that adult female rhesus monkeys tend to show
a higher incidence of affective reactions, data were
gathered on free-living rhesus monkeys on Cayo Santiago.

These data gave no indication that females showed a
higher incidence of affective responses as a whole,
nor that they were more inclined to exhibit the
milder forms of agonistic behavior that Mason et al.
suggested were particularly more prevalent among
females than among males. There was an indication
that the affective social behavior of adult males
was somewhat more likely to be aggressive — that of
females more submissive.

12.   Anderson, C.O., & Mason, W.A.   Early experience and complexity
of social organization in groups of young rhesus monkeys (Macaca
mulatta). Journal of Comparative and Physiological Psychology,
1974, 87(4), 691-690.

Two social groups of young rhesus monkeys, one socially
deprived and one raised with mother and age-mates
showed marked differences in the complexity of social
organization. Compared to deprived monkeys, the
socially experienced animals interacted more often as
trios and larger subgroups their responses with the
triadic subgroups was more complex, and only in this
group did functional aspects of triadic interactions
suggest that one individual recognized the status
relations between other participants. These findings
indicate that higher-order social skills are dependent
on social experience and that these are not necessarily
observable in arranged dyadic encounters.

13.   Anderson, C.O., & Mason, W.A.   Competitive social strategies
in groups of deprived and experienced rhesus monkeys. Developmental
Psychobiology, 1978, 11(4), 289-299.

Behavior during competition for water was observed in
2 social groups of young rhesus monkeys (3 females,
3 males in each). Monkeys in one group were socially
deprived and those in the other were socially experienced
(raised with mother and agemates). Social status,
based on dyadic recording of displacements at the
water bottle, was predictive of a number of measures
related to water consumption and social orientation
in both groups, but this measure was less reliable
and predictive for the experienced group than for the
deprived group. A major reason for the comparatively low
predictive value and reliability of status among

experienced monkeys was their ability to influence
the behavior of higher status members through responses
directed to a 3rd party and other elaborate social
strategies, many of which depended on responding to
status relations between 2nd and 3rd parties. The fact
that such strategies were only observed in the experienced group is a clear indication that the development
of higher orders of social cognition is dependent on
early social experience.

14. Angermeier, W.F., & Phelps, J.B. The effects of differential
early experience upon learning, performance, and biochemical
responses of non-human primates (Rhesus monkeys). 6571st
Aeromedical Research Laboratory Technical Report, ARL-TR-67-10,
Holleman Air Force Base, New Mexico, 1967, 1-64.

A study was made to determine the effects of differential
early experience upon learning, performance, and biochemical correlates of stress. Twenty-eight male rhesus
monkeys from two to four months of age were used for
this research. Four social rearing conditions were
utilized as follows: (1) strict isolation (2) partial
isolation (3) normal laboratory conditions and (4) enriched environment. The study indicated that differential social rearing conditions did not produce differences
of performance in animals subjected to extensive match-
to-sample training under positive and negative reinforcement conditions. However, the results of this exploration
lend support to the hypothesis that biochemical measures
constitute an important supplement to the already
existing wealth of behavioral data involving various
types of stress in nonhuman organisms.

15. Angermeier, W.F., & Phelps, J.B. Early experience and levels
of noxious stimulation in monkeys. Psychologische Forschung,
1971, 34, 246-252.

An analysis conducted in retrospect revealed that monkeys
reared in isolation had a lower required level of shock
(RLS) in order to perform a complex operant task than
did monkeys reared in environments which permitted
varying degrees of visual, tactile and social stimulation.
This behavior prevailed (1) under conditions of both
avoidance and escape training and (2) at the time of
initial exposure to shock and 10 months thereafter.

A limited confirmation of Fuller's theory of emergence (1968) is discussed.

16. Angermeier, W.F., Phelps, J.B., & Reynolds, H.H. The effects of differential early rearing upon discrimination learning in monkeys. Psychonomic Science, 1967, 8(9), 379-380.

Behavioral discrimination tests were performed on 14-16 month-old male rhesus monkeys, reared from the age of 2 to 4 months under four different conditions of perceptual and social enrichment. Results show no differences between any of the groups in the learning of a discrimination task, consisting of two-dimensional four-choice match-to-sample problem, similar in make-up to the PATA-K used by Strong (1965).

17. Angermeier, W.F., Phelps, J.B., & Reynolds, H.H. The relation-ship between performance under stress and blood biochemistry in monkeys. Psychonomic Science, 1967, 8 (9), 389-390.

Twenty-four male rhesus monkeys previously reared under 4 different conditions of social and perceptual deprivation and enrichment were subjected to extensive discrimination training under stress. Biochemical values of the blood, such as Glutamic Oxalacetic Transaminase, Cholinesterase, Tyrosine, Cholesterol, Serum Total Protein, White Blood Count, Albumin, Beta Globulin, and certain Leukocytes seemed to vary virtually point for point with the temporal adaptation to the tasks presented to the animals. The findings were interpreted to indicate that these biochemical measures reflect the degree of performance stress to which the experimental subjects were exposed.

18. Angermeier, W.F., Phelps, J.B., Reynolds, H.H. The effects of differential rearing environments on selected body organs of Macaca mulatta. Areomedical Research Laboratory, Report #ARL-TR-68-8, 1968.

Twenty-six male rhesus monkeys were sacrificed and organ pathology was completed. The following measures were taken: body weight, brain weight, thymus weight, adrenal weight, brain/body ratio, thymus/body ratio,

adrenal/body ratio and gross and microscopic organ
pathology. Differences between these measures based
upon (1) early rearing experience and (2) shock-induced
stress exposure were not significant. Intercorrelations
between these measures were also computed; a number
of them were found to be significant. Performance on
a shock-escape match-to-sample task was found to
correlate significantly with brain weight and brain/
body ratio. Findings of this study do not agree with
results of similar studies conducted on rats.

19. Angermeier, W.F., Phelps, J.B., Murray, S., & Reynolds, H.H.
Dominance in monkeys: Early rearing and home environment.
Psychonomic Science, 1967, 9(7b), 433-434.

Dominance tests were performed on 28-30 month old male
rhesus monkeys, reared from the age of 2-4 months
under four different conditions of perceptual and
social enviornment. Results showed: (1) no differences
between any of the four groups, except those due to
significant differential weight of experimental subjects,
and (2) a "group effect" to be operating which was
somewhat akin to the social relationship among feral
monkeys generally described in the literature as
territory or home range.

20. Angermeier, W.F., Phelps, J.B., Reynolds, H.H., & Davis, R.
Performance and biochemical responses related to social change
vs. chemotherapy in nonhuman primates. Areomedical Research
Laboratory, Report #ARL-TR-67-23, 1967.

A study was made to determine the effects of social
change versus chemotherapy upon performance and biochemical response in nonhuman primates (rhesus monkeys).
Twenty-four male rhesus monkeys from 26-30 months of
age were used for this research. The results indicated
the following: (1) performance of complex discrimination
improves for social subdominant animals changed to
isolation; (2) performance of complex discrimination
shows a decrement for isolated animals which become
subdominant after the change to a state of social
companionship; (3) social status along the dominant-
subdominant scale seems to be more important for prediction
of performance than the perceptual conditions of the
living environment; (4) both changed environments

and injections of Stelazine (trifluoperazine) improved
the biochemical condition of subjects so treated; (5) there
was little or no difference between the relative
therapeutic effects of changed social environments and
Stelazine injections; (6) Stelazine reduced sensitivity
to shock in a shock-escape match-to-sample task
according to degrees of previous environmental stimulation during early rearing; (7) differential rearing
conditions, as used in this study, had no effect upon
any of the factors mentioned above.

21.   Angermeier, W.F., Phelps, J.B., Reynolds, H.H., & Davis, R.L.
Dominance in monkeys: Effects of social change on performance
and biochemistry. Psychonomic Science, 1968, 11(5), 183-184.

Twenty-four differentially reared male rhesus monkeys
were used in this experiment. The animals were tested
on a four choice match-to-sample task. The results
indicated that (1) performance in complex discrimination
improves for social subdominant animals changed to
isolation; (2) performance of the same task shows a
decrement for isolated animals which became subdominant
after a change to a state of social companionship;
(3) control animals and dominant animals were not
affected by social changes; and (4) social status
along the dominant-submissive scale seems to be more
important for prediction of performance than the
perceptual conditions of the living environment.

22.   Angermeier, W.F., Phelps, J.B., Reynolds, H.H., & Davis, R.L.
Early environment and behavioral-biochemical response to trifluoperazine in monkeys. Psychonomic Science, 1968, 11(7),
231-232.

Twenty-four differentially reared male rhesus monkeys
were used as subjects in this experiment. For 30 days
the animals were injected twice daily with trifluoperazine. The effects of these injections were measured on
a match-to-sample task previously learned and on a
number of blood biochemical assessments. Results
indicated that (1) the drug reduced the sensitivity to
shock in a shock-escape match-to-sample task according
to degrees of previous environmental stimulation during
early rearing; (2) the least affected subjects were
the animals reared in strict isolation, animals reared

in partial isolation and under normal social conditions
were moderately affected; (3) these effects could be
interpreted to indicate differential early threshold
development in the four rearing groups used in this
experiment; and (4) drug injections could be considered
"therapeutic" in the same sense that social change was
thought to be therapeutic in a previous study.

23.   Anonymous.   Male monkeys raised only with peers likely to
show sex deficits at maturity.   Primate Record, 1972, 3, 12-13.

This article discusses sexual deficits in peer-reared
male rhesus monkeys. Animals separated from their
mothers and raised only with peers showed immature
mounting behavior and chose females as their mounting
partners much less often than did those monkeys raised
in a situation which provided unrestricted social
contact with mothers and peers.

24.   Arling, G.L., & Harlow, H.F.   Effects of social deprivation on
maternal behavior of rhesus monkeys.   Journal of Comparative and
Physiological Psychology, 1967, 64(3), 371-377.

The primary differences between the design of this
experiment and that of Seay, Alexander, and Harlow (1964)
concerning the maternal behavior of motherless mothers
(MMs) were extending of the observation period and
balancing the groups for sex of the infants. As in
the previous study, females separated at birth from
their own mothers and deprived of early interaction with
peers were distinctly deficient in basic patterns of
maternal behavior when compared with feral-raised mothers.
Also MM infants were deficient in social play and
sexual behavior and hyperaggressive in peer interaction as compared with infants of the feral mothers
and infants in the Seay et al. study.

26.   Arling, G.L., Ruppenthal, G.C., & Mitchell, G.D.   Aggressive
behavior of the eight-year old nulliparous isolate female monkey.
Animal Behavior, 1969, 17, 109-113.

   The infant-directed and age-mate-directed behaviors of
   four 8-year old feral rhesus female monkeys and four 8-
   year-old isolate rhesus females were observed when paired
   in confined quarters with infant and age-mate stimulus
   animals. Although the isolates redirected a smaller
   proportion of their hostility away from the infant than
   did the ferals, they were not abnormally aggressive toward the infants and were certainly much less hostile
   towards infants than were 4-year-old nulliparous female
   isolates described in previous studies. On the other
   hand, the isolate females of the present report directed
   much more physical aggression toward the age-mate stimulus
   animals than did the ferals. As the nulliparous isolate
   female matures, she apparently becomes less aggressive
   toward infants and, if anything, more aggressive toward
   age-mates.

27.   Ausman, L.M., Hayes, K.C., Lage, A., & Hegsted, D.M.   Nursery
care and growth of old and new world infant monkeys.   Laboratory
Animal Care, 1970, 20(5), 907-913.

   In order to use infant simians for studies in experimental
   nutrition, an infant monkey nursery was established.
   Old World (Macaca fascicularis) and New World monkeys
   (Cebus albifrons, Cebus appella, and Saimiri sciureus)
   comprised the experimental animals. The infants were
   removed from the dams at birth, kept in incubators for
   the first 2-3 weeks of life, and then housed in individual
   steel cages. The monkeys were fed a commercial liquid
   baby formula and provided water ad libitum after 3 weeks
   of age. This report includes mean values for growth, food
   consumption, and body temperatures, as well as information
   on clinical care and observations during the first 8
   weeks of life.

28.   Baldwin, D.V., & Suomi, S.J.   Reactions of infant monkeys to
social and nonsocial stimuli.   Folia Primatologica, 1974, 22,
307-314.

   Two studies examined the responses of rhesus monkey infants
   to various stimuli. It was found that all subjects explored

the stimulus objects manually more than orally or pedally
and increased their exploration with increasing exposure
to the stimulus objects. With respect to specific stimuli,
subjects explored a movable object more than a stationary
object, but did not so differentiate between an inanimate
toy squirrel and an animate social peer.

29. Bauer, J., & Held, R. Comparison of visually guided reaching
in normal and deprived infant monkeys. Journal of Experimental
Psychology: Animal Behavior Processes, 1975, 1(14), 298-308.

To assess the deficit of sight-deprived stump tailed
monkeys, a method of open-loop testing (no sight of the
reaching limb) was developed for use on both an experimental
group deprived of sight of limbs, and a control group
raised under identical restraints but allowed sight of
limbs. The results show that the deprived monkeys can
learn to reach a given visible target, but the learned
reach is not as precise in the training condition or as
precise to new target directions as it is for controls.
Furthermore, there is little intermanual transfer of
reaching skill for the experimentals but nearly 100%
transfer for the controls. Finally, experimentals show
a loss of precision in retention testing following a lack
of practice, but controls do not. We conclude that the
differences in visually guided reaching behavior of the
two groups is evidence that the normal accuracy results
from unconstrained vision of the hands which produces a
mapping of the coordinates of motor response onto the
space of vision.

30. Baysinger, C.M. & Suomi, S.J. Effects of periodic loss of
contact comfort on the development of attachment behavior of
rhesus infants. Bulletin of the Psychonomic Society, 1975, 6,
419. (abstract)

Groups of infant rhesus monkeys were exposed to different
schedules of contact comfort provided by mechanical
surrogate mothers. It was found that decreasing the
availability of contact comfort by reducing the surrogate's
surface temperature significantly enhanced subsequent
attachment to the surrogates. The effect was strongest
for infants under 26 weeks of age.

31.   Baysinger, C.M., Brandt, E.M., & Mitchell, G.   Development of infant social isolate monkeys (<u>Mucaca mulatta</u>) in their isolation environments.   <u>Primates</u>, 1972, <u>13</u>(3), 257-270.

This study was designed to more clearly define the treat-ment of early social isolation.   Four newborn infant rhesus monkeys were raised in social isolation and their behavioral development during isolation was compared to the behavioral development of four controls which were reared by mothers.   Isolates spent more time looking at human observers, at the non-social environment, and at themselves then did controls.   They also showed more walking and more manual exploration of the cage and of themselves, but moved more slowly and awkwardly than did the animals raised by their mothers.   Controls were in contact with their mothers much more frequently than isolates were in contact with the towels in their cages. Abnormal movements such as rocking, self-grasping, and autoeroticism appeared in the isolates before the end of the first month.   This suggests that the first month of life is extremely crucial for the development of abnormal behavior.

32.   Baysinger, C.M., Plubell, P.E. & Harlow, H.F.   A variable-temperature surrogate mother for studying attachment in infant monkeys.   <u>Behavior Research Methods and Instrumentation</u>, 1973, <u>5</u>(3), 269-272.

A variable-temperature surrogate mother for use with infant monkeys is described.   The apparatus is designed to facilitate manipulation of the infant-surrogate attachment bond.   Data showing significant behavioral changes in ventral contact and locomotion as a function of depressed surrogate temperature are presented.   The value of this technique in the production of psychopath-ology is indicated by a dramatic and progressive increase in disturbance behaviors during a 90 week test period. Implications for the use of a variable-temperature surro-gate in studying animal models of psychopathology are noted.

33.   Beasley, J.W., & Seal, W.S.   Effects of age and rearing conditions of <u>Macaca mulatta</u> on blood proteins and protein-bound carbohydrates.   <u>Clinical Chemistry</u>, 1968, <u>14</u>(11), 1107-1111.

Measurements of blood constituents were made on monkeys,
(Macaca mulatta), reared in conditions of severe social
isolation which are known to lead to behavioral deficits
in adult life. Alterations were found in the levels of
hemoglobin, total plasma protein, and haptoglobin, while
the levels of fibrinogen and protein-bound carbohydrates
were identical with controls. Differences among the
deprived monkeys did not resemble differences found among
human schizophrenic patients. An effect of age on some
of these measures in the rhesus monkey also was noted.

34. Beckett, P.G.S., Frohman, C.E., Gottlieb, J.S., Mowbray, J.B.,
& Wolf, R.C. Schizophrenic-like mechanisms in monkeys. American
Journal of Psychiatry, 1963, 119, 835-842.

The hypothesis that rhesus monkeys reared in conditions
of social isolation and tactile monotony would show
evidence of a metabolic disturbance which is consistent
with the presence of a blood factor as found in schizophrenic patients was supported, although only a small
group of animals reared under the most extreme conditions
was available, and only these animals, as a group, showed
the abnormality.

35. Benjamin, L.S. The effect of frustration on the nonnutritive
sucking of the infant rhesus monkey. Journal of Comparative and
Physiological Psychology, 1961, 54(6), 700-703.

Five experimental and five control infant rhesus monkeys
were observed from 90 to 165 days of age in a cylindrical
Plexiglas observation cage divided in half by a sliding
Plexiglas partition. The cage was enclosed within a
Masonite cubicle equipped with one-way-vision glass.
Frustration was defined as delay and interference with
the attainment of a preferred food object. Five experimental
monkeys were frustrated for 15 test days by placing preferred
fruit in a dish on the other side of the Plexiglas partition
in the observation cage and restraining them for a 250-sec.
period before they were allowed to eat. Preceding and
following the 15 frustration days were 15 normative and
15 extinction days, respectively. Control animals never
received food in the experimental situation. Higher
nonnutritive sucking scores were made by the five experimental Ss than by the five control Ss during the frustration test period, and these differences were statistically

significant. Differences between these two groups of
S̲s̲ during normative and extinction periods were not
significant.

36. Berger, R.J., & Meier, G.W.   The effects of selective deprivation of states of sleep in the developing monkey. Psychophysiology,
1966, 2̲(4), 354-371.

     Four groups, two comprising three neonatal rhesus monkeys
     and two comprising two juvenile rhesus monkeys, were
     selectively deprived of either low-voltage, fast-wave sleep
     (LVF) or of high-voltage, slow-wave sleep (HVS), respectively.
     Both infant and juvenile S̲s̲ displayed an over-all increase
     in threshold to the tone-shock combination during the
     deprivation of either phase of sleep. However, the
     thresholds of the infant S̲s̲ were greater, throughout
     deprivation, than the thresholds of the juvenile S̲s̲. The
     juvenile S̲s̲ exposed to LVF deprivation were unique in
     exhibiting a sharp increase in frequency of forced awaken-
     ings from LVF, to values significantly greater than for
     the other groups, and in displaying compensatory recovery
     effects, manifested by increases in proportion of total
     sleep time spent in LVF, following termination of depri-
     vation.

37. Berger, R.J., & Meier, G.W.   Eye movement during sleep and
waking in infant monkeys (Macaca mulatta) deprived of patterned
vision. Developmental Psychobiology, 1969, 1(4), 266-275.

     Five newborn rhesus monkeys deprived of patterned vision
     and 4 normal-sighted controls were reared until 18 months
     of age. There was a significant decrease in frequency
     of rapid eye movement (REM) in both groups with increasing
     age. Frequency of REM tended to be lower in experimental
     monkeys than in controls. Asymmetries in velocity of
     waking, horizontal eye movement were evident in experi-
     mental monkeys, and the velocity of REMs was less in
     pattern-deprived monkeys than in controls. Nonsequential
     interval histograms between REMs at 18 months of age
     differed significantly from a random exponential distri-
     bution in both groups. Pattern-deprived monkeys had
     significantly fewer short intervals between REMs than
     controls. Visual tests showed the pattern-deprived monkeys
     to be behaviorally blind on removal of their eye occluders
     at 18 months of age.

38. Berkson, G. Abnormal stereotyped motor acts. In J. Zubin &
H. Hunt (Eds.), Comparative psychopathology. New York: Grune &
Stratton, 1967.

   The purpose of this paper is to describe some features of
   stereotyped motor acts common to both lower primates and
   humans. A number of experiments were performed with chimpanzees and mental defectives to determine the kinds of
   variables which affect changes in the level of stereotyped
   behaviors once they have developed. It was found that in
   chimpanzees, rocking or swaying is determined by both the
   general arousal level of the animal and by the extent to
   which the environment elicits activities other than the
   stereotyped behaviors. With defectives, the main variable
   seemed to be the extent to which the person responds to
   the environment.

39. Berkson, G. Development of abnormal stereotyped behaviors.
Developmental Psychobiology, 1968, 1(2), 118-132.

   Groups of five crabeating macaques were separated from
   their mothers at 0, 1, 2, 4, or 6 months of age and
   observed during their first year of life. Abnormal
   stereotyped behaviors developed in all groups, but the
   frequencies of different patterns were modified by age at
   isolation. Self-sucking, crouching, and self-grasping
   developed almost immediately after birth, while body rocking
   and repetitive locomotion began later. Animals separated
   at a later age were more active and aggressive in a
   novel environment than were those separated early.
   Social development in the first 6 months was similar to
   that of other macaques.

40. Berkson, G. Defective infants in a feral monkey group. Folia
Primatologica,1970, 12, 284-289.

   Subhuman primates can survive severe injuries in a natural
   habitat. In this study two infant monkeys in a natural
   group were made partially blind and were observed to
   determine whether they would survive and, if so, to take
   an initial look at their status in the group. The infants
   lived until they were seven months old and during that
   time spent a good bit of time separate from their mothers.
   Instances of group responses to stress and of intragroup
   relationships with special reference to the defective infants
   are described.

41.   Berkson, G.   Social responses to abnormal infant monkeys.
American Journal of Physical Anthropology,   1973, 38, 583-586.

    Experimental modification of an individual's behavior
    can be used to evaluate the range of group social reactions
    to him.   This principle is used in evaluating the degree
    to which macaque groups compensate for an infant's
    clumsiness, consequent on reduced visual acuity from an
    experimental cataract.   Evidence is presented to support
    the following propositions: Defective infants are not
    killed by the group even in crowded conditions; compen-
    satory care is given during the first year, primarily
    by the mother and to a certain extent by other animals
    in the group; isolation of the infant begins in the
    second year; social responses to the defective infant
    reflect normal social responses.   Compensatory care of
    chronically defective individuals is a consequence of
    evolution of social behaviors adapted to nurture young
    individuals over a long period of dependence and to protect
    other group members during temporary periods of weakness.

42.   Berkson, G.   Visual defect does not produce stereotyped
movements.   American Journal of Mental Deficiency, 1973, 78(1),
89-94.

    Monkeys with a severe visual acuity deficit were reared
    in social isolation. They developed abnormal stereotyped
    movements but at no greater rate than sighted isolation-
    reared controls.   A consideration of the recent literature
    suggests that specific deficits in social stimulation
    are implicated in the development of various stereotyped
    acts.

43.   Berkson, G.   Social responses of animals to infants with defects.
In   M. Lewis and L. Rosenblum (Eds.), The effect of the infant on
its caregiver.   N.Y.:   Wiley, 1974.

    Modern evolutionary theory recognizes that the survival
    of many species of animals has been dependent on their
    ability to develop complex societies in which individuals
    cooperate with one another to acquire food, rear young,
    and defend against predators.   This chapter proposes that
    these kinds of cooperation contribute to the survival of
    abnormal individuals, and that the presence of abnormal
    animals provides a way of analyzing the physical and

social ecology of animal groups. A number of studies
with crabeating monkeys (Macaca fascicularis) mothers and
their blind infants in both laboratory and free-ranging
environments are described.

44. Berkson, G. Social responses to blind infant monkeys. In
N.R. Ellis (Ed.), Abberant development in infancy. Hillsdale N.J.:
Erlbaum Assoc., 1975.

    Six partially blind and 5 totally blind infant monkeys
    (Macaca fascicularis) were observed in 4 laboratory groups
    for a period of one year. No evidence of social selection
    for visual deficits was found, and although compensatory
    care seemed to be the rule, it appeared to depend on the
    extent to which the environment made the visual defect
    obvious.

45.Berkson, G. Rejection of abnormal strangers from macaque monkey
groups. Journal of Abnormal Psychology, 1977, 86(6), 659-661.

    Host group attacks on monkeys (Macaca fascicularis) that
    were strangers were not affected by the strangers' being
    blind or sedated. Exploration by the host group of
    these strange monkeys depended on the novelty of the
    strangers' behavioral state.

46. Berkson, G. The social ecology of defects in primates. In
S. Chevalier-Skolnikoff and F.E. Poirier (Eds.), Primate biosocial
development: Biological social and ecological determinants. New
York: Garland, 1977, 189-204.

    This paper considers the relationship between aspects of
    higher primate social organization and response of the
    handicapped. The results of studies of macaque monkey
    responses to visually impaired members of their group
    are presented and the results discussed in relation to
    an ecological concept of handicap.

47. Berkson, G., and Becker, J.D. Facial expressions and social
responsiveness of blind monkeys. Journal of Abnormal Psychology,
1975, 84(5), 519-523.

Three studies show that tactual and perhaps auditory cues
maintain social affinity of blind juvenile macaque monkeys.
Olfactory discrimination of individuals was tested but
not demonstrated. Facial expressions are essentially
normal in form, but in some circumstances they occur at
different frequencies than those of sighted animals.

48. Berkson, G. & Karrer, R. Travel vision in infant monkeys:
Maturation rate and abnormal stereotyped behavior. Developmental
Psychobiology, 1968, 1(3), 170-174.

Three monkeys with travel vision and 5 controls were
observed with their mothers during the first 6 months
of age and then for another 3 months after they had
been placed in social isolation. In the home cage, the
experimental animals did not differ from controls except
that two held a hand before their eyes in a stereotyped
fashion not previously reported for animals. In an
unfamiliar environment the blind animals did not look
at an observer, threatened less than normal, and were
awkward in moving around. Animals who held hands before
eyes tended to approach a flickering visual stimulus.

49. Berkson, G. and Mason, W.A. Stereotyped behaviors of chimpanzees: Relation to general arousal and alternative activities.
Perceptual and Motor Skills, 1964, 19, 935-652.

Chimpanzees raised without their mothers develop abnormal
stereotyped behaviors not found in mother-raised animals.
Two studies tested the hypotheses that moment-to-moment
fluctuations of stereotyped behaviors are related to
the level of general arousal and to the extent to which
alternative activities are performed. The first study
showed that body rocking or swaying is positively related
to white noise level and to food deprivation. In the
second experiment, rocking and swaying decreased during
habituation to a novel situation and returned to a high
level following administration of amphetamine. Activities
alternative to stereotyped movements were evoked in a
third experiment and produced a decrement in rocking and
swaying. In general, the evidence suggests a positive
relationship between arousal and rocking and swaying;
the relationship of arousal to other stereotyped acts
was not as regular. No behavior in the repertoire of
mother-raised animals was found to be homologous to

Rock-Sway. The relationship between the various behavior
categories studied was related to S's rearing history,
momentary arousal level, and the extent to which the
cues in the situation evoked the various behaviors.

50. Berkson, G., Goodrich, J., & Kraft, I. Abnormal stereotyped
movements of marmosets. Perceptual and Motor Skills, 1966, 23, 491-498.

Observations of a laboratory marmoset colony suggest
that marmosets reared in isolation do not develop the
abnormal stereotyped behaviors found in isolation-related
macaques and chimpanzees. However, juvenile marmosets
can develop "cage stereotypes".

51. Berkson, G., Mason, W.A., & Saxon, S.V. Situation and stimulus
effects on stereotyped behaviors of chimpanzees. Journal of
Comparative and Physiological Psychology, 1963, 56(4), 786-792.

Six laboratory raised chimpanzees were tested in 4 experiments
designed to analyze stimulus and situational factors affecting stereotyped behaviors characteristic of primates
raised without their mothers. The level of stereotyped
behaviors was highest in an enclosed cubicle. This
effect was apparently not related to cage size nor was
it importantly related to novelty of the situations.
Opportunity to manipulate objects decreased stereotyped
responses, and changes in frequency of stereotyped
behaviors were accompanied by changes in other behaviors.
Fear producing stimuli increased the level of repetitive
stereotyped movements although these movements also
occurred when such stimuli were not present. It was
suggested that stereotyped behaviors are directly associated
with general level of arousal and that they are reduced
when alternative activities are evoked.

52. Berman, A.J., Waizer, J., Dalton, L. Jr. Consequences of
asphyxia at birth in the monkey. Medical Primatology, 1970
Proceedings of the 2nd Conference of Experimental Medical Surgery
in Primates New York, 1969, (Karger/Basel, 1971)

Three infant rhesus monkeys (Macaca mulatta) asphyxiated
at birth and subsequently resuscitated were found to be
slower to reach criterion on 3 different schedules of

reinforcement than were normal monkeys. In addition,
2 of the 3 asphyxiated subjects were unable to maintain
a steady rate of response on a variable-ratio schedule,
while all three of the controls were able to achieve a
high rate of response. The experimental animals also
displayed a large number of abberant behavior patterns
while in their individual home cages which were not
noted to the same extent in normal monkeys similarly
housed.

53. Bernstein, I.S. Group social patterns as influenced by removal
and later reintroduction of the dominant male rhesus. Psychological
Reports, 1964, 14, 3-10.

The dominant male in a rhesus monkey group was removed
from the group and returned after a period of one month.
During the period of his removal the social activities
of the remaining males increased. Upon return the dominant
male assumed his former position showing increased social
activity whereas the activities of other males were reduced.

54. Bernstein, S. & Mason, W.A. The effects of age and stimulus
conditions on the emotional responses of rhesus monkeys: Responses
to complex stimuli. Journal of Genetic Psychology, 1962, 101,
279-298.

This experiment investigated the responses of 47 rhesus
monkeys ranging in age from one to 25 months, to 12 stimulus
objects arranged in three levels of complexity (simple,
intermediate and complex). The recorded behaviors included
ten responses generally assumed to be indicative of
emotional distress in the rhesus monkey, and five responses
relating to orientation in space. The first ten responses
were combined as a total emotion score. In all age groups
the total emotion score increased from the simple to the
complex stimulus conditions. Examination of individual
response characteristics suggested the following trends
in the development of emotional responsiveness in the
laboratory-reared rhesus monkey. From birth to three
months of age emotional behavior consists principally
of vocalization and nondirected responses such as rocking,
crouching and sucking. Between three months and two
years of age there is a rapid and progressive increase
in the frequency of directed responses including barking,
lipsmacking, drawing back the ears, and the fear grimace.

BIBLIOGRAPHY   23

Concurrent with these changes there is an increase in
the tendency to withdraw from a disturbing stimulus.
These changes in the patterns of emotional responses are
probably correlated in the native habitat with the
waning of the primary mother-infant bond and the corresponding growth of filial independence.

55. Bernstein, S. & Mason, W.A. Effects of age and stimulus
conditions on the emotional responses of rhesus monkeys: Differential responses to frustration and to fear stimuli. Developmental
Psychobiology, 1970, 3(1), 5-12.

Laboratory-raised rhesus monkeys ranging in age from 3
to 25 months were tested for reactions to toys representing
animals ('fear stimuli') and to food frustration. The
results indicate that these test conditions are associated
with different response profiles and that the contrast
between responses to fear and to frustration increases
with age. Frustration in older animals is associated
with regression to a less mature level of functioning.

56. Bernstein, I.S., Gordon, T.P., & Rose, R.M. Aggression and
social controls in rhesus monkey (Macaca mulatta) groups revealed
in group formation studies. Folia Primatologica, 1974, 21, 81-107.

Agonistic responses were the primary form of social
interaction during five rhesus group formations. Males
showed the most extreme forms of aggression initially,
but as formative processes progressed, females became
more active and aggression less severe. Initial agonistic
interactions serve to establish social order of an
emerging group. As a group becomes organized, extreme
forms of aggression disappear and aggressive frequencies
decline.

57. Bernstein, I.S., Gordon, T.P., & Rose, R.M. Factors influencing the expression of aggression during introductions to rhesus
monkeys groups.   In R.L. Holloway (Ed.), Primate aggression,
territorality, and xenophobia. New York: Academic Press, 1974.

A series of over 75 introductions of rhesus monkeys
(Macaca mulatta) into established groups was conducted
over a 2-year period at the Yerkes Field Station. It

was hypothesized that the age and sex of the animals
being introduced, the sex composition of the group the
animal is being introduced to, the previous social history
of the introduced animal with that group, and the individual response patterns of animals introduced to a group,
would all influence the amount and type of aggressive
responses directed to new-comers by a host group. Although
the series of experiments reported provided data relevant
to some of the proposed hypotheses, because of the
variability of the results, the authors warn against the
adoption of a simplistic model of aggression in lower
primates and caution against wide generalizations which
go beyond the constraints of the supporting evidence.

58. Blomquist, A.J., & Harlow, H.F. The infant rhesus monkey
program at the University of Wisconsin Primate Laboratory.
Proceedings of the Animal Care Panel, 1961, 11(2), 57-64.

      Over a period of five years, adult rhesus monkeys were
      mated at the Wisconsin Primate Laboratory. The infants
      were raised apart from their mothers beginning with day
      one of life. A detailed description of the procedures
      developed both in adult mating and in infant care over
      this five-year period is given.

59. Blomquist, A.J., Harlow, H.F., Schiltz, K.A., & Mohr, D.
The effect of temporal lobe lesions on discrimination learning
performance by infant, adolescent and adult rhesus monkeys.
The Gerontologist, 1971, 11(3), 41.

      The cortex of the inferior temporal lobe was removed
      bilaterally under sterile conditions in four 100 day
      old rhesus monkeys, four 300 day old rhesus monkeys and
      four 730 day old rhesus monkeys. For each of the three
      groups of brain damaged monkeys, a group of four monkeys
      of comparable age was used as a control. Following
      recovery from surgery by the experimental subjects, all
      animals were tested with object and pattern discrimin-
      ation problems. The results showed no significant
      differences in performance between the experimental and
      control animals for the 100 and 300 day old groups.
      In contrast, the older, 730 day old experimental animals
      were significantly inferior to the 730 day old control
      animals in performance on the pattern discrimination

problems.

60. Bobbitt, R.A., Jensen, G.D., And Gordon, B.N.   Behavioral
elements (taxonomy) for observing mother-infant-peer interaction
in Macaca nemestrina.   Primates, 1964 5(3-4), 71-80.

   The authors present a rather detailed behavioral taxonomy
   developed for observing mother and infant pigtail macaques.
   This taxonomy has been used according to an observational
   system developed in this laboratory and employed in
   several specific studies.   It has been determined
   empirically that observers are able to maintain a satis-
   factory level of reliability.

61. Boothe, R., Teller, D.Y., & Sackett, G.P.   Trichromacy in
normally reared and light deprived infant monkeys (Macaca
nemestrina).   Vision Research, 1975, 15, 1187-1191.

   Infant macaque monkeys (Macaca nemestrina) were separated
   from their mothers a few days after birth and individually
   housed in a specially designed testing cage.   Each
   was taught to discriminate white light from each of
   several narrow-band wavelengths of light selected from
   across the visible spectrum.   During the first 2 months
   after birth, each of the infants learned to discriminate
   all wavelengths tested from white light, regardless of
   relative luminance.   An infant which had been raised in
   continuous darkness from 2 weeks until 3 months after
   birth was similarly tested.   This dark-reared infant
   also successfully learned to discriminate all wavelengths tested from white light.   It is concluded that
   infant pigtail monkeys have trichromatic color vision
   by the age of 2 months and that their trichromacy remains
   present following a period of dark rearing during the
   first 3 months after birth.

62. Bowden, D.M., & McKinney, W.T.   Behavioral effects of peer
separation, isolation, and reunion on adolescent male rhesus monkeys.
Developmental Psychobiology, 1972, 5(4), 353-362.

   Six adolescent male rhesus monkeys which had lived in
   pairs for 6-8 months were studied for behavioral adaptation

to 2 weeks of separation-isolation and 4 days of reunion.
The response to separation was 1-3 days of increased
locomotion and object-oriented behavior followed by a
long term increase in self-directed behaviors. The
latter phase differed significantly from the anaclitic
depression of infants separated from their mothers.
Reunion produced one day of intense social activity
followed by a return to behavior patterns characteristic
of the preseparation period. A large component of the
initial motor excitement following separation may be
nonspecific response to gross change in the environment.

63. Bowden, D.M. & McKinney, W.T. Jr. Effects of selective frontal
lobe lesions on response to separation in adolescent rhesus monkeys.
Brain Research, 1974, 75, 167-171.

The results of the present study suggest a possible neural
mechanism contributing to the difference between infant
and adolescent responses to separation and contribute
further support for a limited hypothesis regarding the
role of the frontal lobes in the control of gross activity
level. Adolescent animals with frontal lobe lesions
imposed after infancy showed a hypoactive response to
separation-isolation which adolescents with intact frontal
lobes do not show. Since the response consisted of a
hypoactive phase in adaptation, the concept that the
frontal lobes play primarily an inhibitory role in
modulating arousal is inadequate to explain the phenomenon;
the concept that they limit fluctuations in activity level
at both extremes may provide a more adequate interpretation
of the data.

64.   Bowden, D.M., Goldman, P.S., Rosvold, H.E., & Greenstreet, R.L.
Free behavior of rhesus monkeys following lesions of the dorso-
lateral and orbital prefrontal cortex in infancy. Experimental
Brain Research, 1971, 12, 265-274.

The present study sought to determine whether differential
effects could be found in the free behavior of early-
operated monkeys with selective removal of the dorsolateral
and orbital prefrontal cortex. The early-operated
monkeys were first observed in their home cages in stable
living groups which had existed for at least 6½ months
(together condition) and then again immediately after
they had been separated from one another (separated

conditions). The major results indicated that the monkeys
with orbital lesions differed from unoperated controls
in more ways than did the dorsolateral monkeys: in the
together condition, the orbital group slept more and were
more sedentary even when awake; in the separated condition,
by contrast, they became hyperactive and spent most of
their time in locomotion. These findings lend support
to the notion, derived from studies of cognitive behavior
in infant-operated monkeys, that functions of the orbital
cortex, unlike those of the dorsolateral cortex, are not
spared following brain damage in infancy. Further, the
nature of the behavior exhibited by the impaired monkeys
led to the suggestion that the orbital cortex may play
an important role in modulating arousal mechanisms in the
infant and adult monkey alike.

65. Bowlby, J. Separation anxiety. International Journal of
Psychoanalysis, 1960, 41, 89-113.

Theoretical issues raised by the observation of young
children 15-20 months old separated from their mothers
and cared for by strangers are discussed. Theoretical
problems are examined with reference to three observed
phases of behavior: The phase of Protest which raises
the problem especially of separation anxiety; Despair,
that of grief and mourning; and Detachment, that of
defense. The thesis advanced by the author is that the
3 types of responses-separation anxiety, grief and
mourning, and defense--are phases of a single process
and that, when treated as such, each illumines the
other two.

66. Brandt, E.M., Baysinger, C., & Mitchell, G. Separation from
rearing environment in mother-reared and isolation-reared rhesus
monkeys (Macaca mulatta). International Journal of Psychobiology,
1972, 2(3), 193-204.

The study compared the effects of two extremely different
preseparation rearing environments on the behaviors of
infant rhesus monkeys during separation from these rearing
environments. Comparisons between isolation-reared and
mother-reared monkeys were made before, during, and after
early separation from rearing environment and mother.
Stereotyped movements appeared in the control infants
in response to separation, while isolate stereotype

occurred before and after as well as during separation.
Both isolates and controls showed increased self-directed
behaviors during separation. Vocalization specifically
associated with separation in the controls was used
less often and outside of the separation stage by the
isolates. It was concluded that the nature of the rearing
environment prior to separation has effects on the change
in behaviors typically seen when animals are separated
from their objects of attachment. This study also
compared the effects of separation on mothers as opposed
to the effects of separation on infants. Mothers
responded to separation with facial expressions, infants
with vocalizations.

67. Brandt, E.M., & Mitchell, G.D. Pairing preadolescents with
infants (Macaca mulatta). Developmental Psychology, 1973, 8(2),
222-228.

This research was concerned with three topics: (a) pre-
parental behavior in male and female rhesus monkeys,
(b) sex differences in infant monkeys, and (c) the effects
of early social isolation. Each of eight male or female
isolate or control infants was paired for 3 weeks with
either a male or a female preadolescent. The male
preadolescents were more aggressive toward the infants
than were the females, who were at times very gentle.
The males did show some affection for the infants,
especially female infants, but it was awkward and often
rough. The male infants both elicited and emitted
more aggressive behavior in their pairings with preadolescents than did the female infants. However, they
also elicited and emitted play. There was more social
behavior directed toward control infants than toward
isolate-reared infants, but social interaction between
preadolescents, particularly females, and the isolate
infants increased significantly over time. Normally
socialized female preadolescents may provide good social
therapy for isolate-reared infant rhesus monkeys.

68. Brandt, E.M., Stevens, C.W., & Mitchell, G. Visual social
communication in adult male isolate-reared monkeys (Macaca mulatta).
Primates, 1971, 12(2), 105-112.

The social responses of isolate-reared and control adult
males were compared in a situation where each subject

was allowed visual and auditory contact but no physical
contact with two different stimulus animals. The isolates
displayed significantly more self-directed hostility and
significantly more masturbation than did controls.
Isolates did not habituate or adapt to the presence of
an adult stimulus animal as readily as did the controls.
A juvenile stimulus animal vocalized significantly more
frequently in the presence of an isolate than she did
in the presence of a control male. Visual (facial,
postural) and auditory (vocal) communication in adult
isolate-reared monkeys is abnormal even when no physical
contact is permitted between the isolate and another animal.

69.   Broadhurst, P.L.   Abnormal animal behavior.   In H.J. Eysenck,
(Ed.), <u>Handbook of abnormal psychology</u>. New York: Basic Books, 1961.

    Since the earliest experiments involving experimental
    neurosis in Pavlov's laboratory before World War I, a
    body of work on this topic has accumulated, much of
    which purports to be relevent to human psychopathology.
    This chapter subjects this literature to scrutiny in
    order to assess its worth and its relevance to the
    human field. Abnormalities of behavior produced in
    infrahuman animals by psychological or functional means
    as well as methods which have been used in attempted
    therapy of abnormal behavior are discussed.

70.   Bronfenbrenner, U.   Early deprivation in mammals: A cross-
species analysis. In   G. Newton and S. Levine (Eds.), <u>Early
experience and behavior</u>. Springfield Ill.:   Thomas, 1968,
627-764.

    This chapter examines the effects of early deprivation
    across species with the aim of identifying continuities
    and gradients as one ascends the phylogenetic ladder.
    This cross-species comparison is confined to studies of
    mammals, and concentrates on rodents, cats, dogs, and
    monkeys. Twenty hypotheses derived from research on
    lower mammals are examined with respect to their
    applicability to man.

71. Butler, R.A. The effect of deprivation of visual incentives
on visual exploration motivation in monkeys. Journal of Physiological
Psychology, 1957, 50, 177-179.

    Five rhesus monkeys were deprived of a varied visual
    experience for 0, 2, 4, and 8 hrs. The animals were
    then tested to determine whether their responses to
    visual incentives would be increased as the result of
    this deprivation. A variable-interval reinforcement
    schedule was used to test the motivational strength of
    the monkeys. Reinforcement consisted of a 12-sec. view
    of the monkey colony outside the test cage. The results
    showed that the number of responses to visual incentives
    approached a maximum after 4 hrs. of deprivation.
    Differences between mean response frequencies for the
    various conditions were significant at the .05 confidence
    level. These data provide another demonstration of
    the similarities between behavior based on a proposed
    curiosity motive and behavior based on biological drives.

72. Candland, D.K., & Mason, W.A. Infant monkey heart rate:
Habituation and effects of social substitutes. Developmental
Psychobiology, 1969, 1(4), 254-256.

    The heart rates of four infant monkeys that habitually
    clung to terry-cloth towels were measured by frequency
    modulation (FM) telemetry while the animals were (1)
    alone in an unfamiliar room, (2) in a second unfamiliar
    room with a towel available, and (3) in the second room
    without a towel. The absence of the towel produced the
    lowest heart rate. Under all conditions, heart rate
    decreased within sessions and over repeated exposures.

73. Castell, R. & Sackett, G.P. Motor behaviors of neonatal rhesus
monkeys: Measurement techniques and early development. Developmental
Psychobiology, 1973, 6(3), 191-202.

    Development of newborn monkeys was assessed by 2 techniques.
    The 1st was a film analysis measuring aspects of walking,
    climbing, and jumping in mother-reared infants. The 2nd
    used a "Baby-Spin" apparatus consisting of a cloth covered
    bag which rotated in the vertical plane through 360°, 2 sec.
    spins. Mother-reared monkeys walked and climbed on Week
    1 but did not jump until Week 3. The duration of arm
    and leg movements was constant, but because distance of

movements changed the speed of limb movement increased
over Weeks 1-5. Clasping and grasping responses were in
synchrony from the first week of life. Clasping pressure
to initial acceleration increased over the first 3 weeks,
with greatest pressures at head-down angles on Weeks 1
and 2, and a double-peaked function, appearing on Weeks
3 and 4 with pressure maxima at both head-down and head-up
angles. Developmental changes in clasping could be seen
in longitudinal or cross-sectional comparisons, or in
comparisons confounded by effects of prior test experience.

74.   Castell, R., & Wilson, C.   Influence of environment on
development of mother-infant interaction in pigtail monkeys.
Proceedings of the 3rd International Congress of Primatology,
Zurich, 1970, 3, 121-124.   (Karger/Basel)

    This study examines the behavioral aspects of spatial
    and social restrictions on developing infant pigtail
    monkeys in groups with their mothers and individually
    caged mother-infant pairs. The mothers of the group-
    living pairs were much more protective of their infants
    in the first 8 wks. of life than were the mothers of the
    individually caged pairs. During the later months, the
    time spent away from mother by group-reared infants
    was far greater and may reflect a socialization process
    possible only in a group situation.

75.   Chalmers, N.R.   Changes in mother-infant behavior following
changes in group composition in sykes and patas monkeys.   Proceedings of the 3rd International Congress of Primatology, Zurich,
1970, 3, 116-120 (Karger/Basel)

    The influence of adult males on the behaviour of mothers
    towards their infants was investigated in groups of
    captive Sykes and Patas monkeys. Each group consisted
    of 1 adult male, 2 to 5 adult females and their young.
    Removal of the adult male caused a decrease in the time
    spent by the infant away from the mother in Sykes but
    not in Patas. It also caused an increase in the mother's
    contribution towards maintaining proximity between herself
    and her infant, and in the amount of time spent holding
    the infant. Removal of an adult female did not produce
    such changes in behaviour. Similar changes in group
    composition did not produce comparable changes in Patas
    monkeys. An explanation for this difference between the
    species is suggested.

76.   Chamove, A.S.   Rearing infant rhesus together.   Behavior, 1973, 47(1-2), 48-66.

    To analyze the antecedent conditions of the together-
    together syndrome, 6 subjects were reared in pairs, 6
    in pairs separated on alternate weeks, 6 in pairs which
    changed in composition weekly, 6 in a group of 6, and 6
    in individual cages. When tested in groups of 6 during
    the first year of life and with infants, juveniles, and
    adults during the second year of life, it was found that:
    (a) self-play and social play increased and self-cling
    and aggression decreased as an increasing function of the
    number of rearing partners; (b) social cling was greatest
    in subjects reared constantly with the same animals, and
    was independent of the amount of play exhibited by a group.

77.   Chamove, A.S.   Social behavior comparison in laboratory-reared
stumptail and rhesus macaques.   Folia Primatologica, 1973, 19, 35-40.

    No differences in social behavior during the first year
    of life were found in comparisons of two quadrads of each
    macaque type when controlling for adult-related differences
    by rearing infants with one another. No support was found
    for reports of feral stumptails being more affiliative
    or less assertive. Another group of rhesus tested in
    a larger enclosure were more assertive and more fearful
    when compared with rhesus in a small test cage.

78.   Chamove, A.S.   Varying infant rhesus social housing.   Journal
of the Institute of Animal Technicians, 1973, 24(1), 5-15.

    A socially adequate rhesus monkey can be produced from
    housing with relatively restricted social opportunity.
    Social experience with more than one other peer, or age-
    mate, from around the age of three months either (a)
    housed continuously in a group, (b) housed continuously
    in pairs which change weekly, (c) given 30 minutes of
    social experience in changing pairs in a rather small
    cage-either of these (or perhaps even better some combination)
    results in animals having a level of social facility desirable
    in subsequently individual-or group-housed macaques.

79. Chamove, A.S. Failure to find rhesus observational learning.
Journal of Behavioral Science, 1974, 2(2), 29-41.

No improvement was found on any trial of 40 learning
set problems using 8 group-living macaques as both
demonstrators and observers, and varying degrees of
sophistication and amount of observation opportunity
when given opportunity to observe others solving the
same problems.

80. Chamove, A.S. Deprivation of vision in social interaction in
monkeys. Journal of Visual Impairment and Blindness, 1978, 72 103.

In order to delineate the role of vision in social development, stumptailed macaques (Macaca archtoides) were
separated from their mothers and given all of their social
experience in the dark. Preliminary results suggest
that (a) seeing others during social interaction is not
essential for adequate social development; (b) the
balance of assertive behaviors are altered by the absence
of visual social stimuli; (c) the absence of social visual
cues, usually releasing stimuli for aggression, alter the
expression of aggression both in short and long term
in a way unexplained by current theories of aggression;
(d) the role of vision in social development may be more
specific than that suggested by the literature.

81. Chamove, A.S. Therapy of isolate rhesus: Different partners
and social behavior. Child Development, 1978, 49, 43-50.

Thirty-six monkeys from 3 different rearing conditions were
subdivided into 3 groups and then paired daily for 20
weeks either with others that were (a) socially sophisticated
9-month old monkeys; (b) partial isolates, reared alone
from birth for 9 months; or (c) socially naive 3-month-
old infant monkeys. When 12 9-month-old partial isolates
were thus split and tested, social play was greatest
in isolates paired with sophisticated therapists and
least in isolates paired with isolate therapists. Fear
showed the reverse pattern. Hostility was greatest
in those infants, isolates, and sophisticates when paired
with isolates and least in all groups when paired with
infants. Of the nonisolate groups the infants were
affected by their pairing the most, showing reduced social

play when paired with the isolates, but increasing play
when with sophisticates. Evidence is found against a
learning or instrumental model of aggression production,
and a novel theory is advanced suggesting that infants
are genetically predisposed to acquire specific behavioral
characteristics shown by the mother.

82.   Chamove, A.S., & Molinaro, T.J.   Monkey retardate learning
analysis.   Journal of Mental Deficiency Research, 1978, 22, 37-48.

    Seven rhesus monkeys reared on diets high in phenylalanine
    (PKU), were compared with normal controls, pair-fed
    controls, younger controls, frontal brain-lesioned monkeys,
    and those raised on another high amino acid diet, tryptophan.
    Error-types in the PKU group, unlike any of the comparison
    groups, were revealed after detailed analysis of three
    object discrimination tasks, viz. learning-set, successive-
    discrimination, and a task with two objects both of which
    were baited. The error patterns suggest that following
    nonreward, indicating an error, the PKU group does not
    shift its response to other objects as readily as the
    control group when this lose-shift strategy is an optimal
    one. This is compatible with an emotionality inter-
    pretation of the PKU learning deficit.

83.   Chamove, A.S., Eysenck, H.J., & Harlow, H.F.   Personality in
monkeys: Factor analyses of rhesus social behavior.   Quarterly
Journal of Experimental Psychology, 1972, 24, 496-504.

    Three factor analyses were performed on social interaction
    data from 168 juvenile macaques. Animals were tested
    in stable quadrad peer groups; in newly-formed dyads
    with infant, juvenile, and adult stimulus monkeys; and
    in similar triads with the stimulus animal plus a familiar
    cage-mate. Factors emerged, most strongly in the most
    stable condition, which were interpreted as affiliative,
    hostile and fearful. These factors were almost entirely
    independent and resembled the extraversion, psychoticism,
    and emotionality factors frequently found in humans.

84.   Chamove, A.S., Harlow, H.F., & Mitchell, G.   Sex differences
in the infant directed behavior of preadolescent rhesus monkeys.
Child Development, 1967, 38, 329-335.

Fifteen pairs of preadolescent rhesus monkeys matched for
rearing experience were tested with a 1-month-old infant.
Preadolescent females directed significantly more positive
social behavior (p < .05) and significantly less hostility
(p < .01) toward the infant than did males. These results
are taken as evidence that hormonal changes at puberty
are not the only variables producing sex differences in
infant-directed behavior.

85. Chamove, A.S., Kerr, G.R., & Harlow, H.F. Learning in monkeys
fed elevated amino acid diets. Journal of Medical Primatology,
1973, 2, 223-235.

Diets containing altered levels of specific amino acids
or their metabolites were fed to rhesus macaques during
infancy and during fetal life. Learning tests, after
the animals were placed on a normal diet, indicated
permanent mental retardation in monkeys that received
phenylalanine either prenatally or for 3, 6, or 12 months
postnatally, and also in monkeys fed parachlorophenylalanine,
which inhibits the metabolism of phenylalanine. Retardation
was not detected in monkeys fed diets high in other amino
acids or in pair-fed control animals.

86. Chamove, A.S., Rosenblum, L.A., & Harlow, H.F. Monkeys (Macaca
mulatta) raised only with peers: A pilot study. Animal Behavior,
1973, 21, 316-325.

Four infant rhesus raised in a group (4-TT) and six raised
in pairs (2-TT) were compared with eight infants raised
on mother surrogates (SP) and twenty raised with real mothers
(MP). When tested with peers early in life 4-TT and 2-TT
subjects showed less play, hostility and sex, and the
2-TT subjects exhibited a preponderance of social cling.
When tested as adults the 2-TT and 4-TT monkeys were
below controls on measures of play, above controls on
social proximity, hostility, and withdrawal, and the
2-TT subjects showed inadequate sexual adjustment. Data
are interpreted as suggesting that behaviours normally
associated with affectional ties can become so extreme
as to inhibit normal social development.

87.   Chamove, A.S., Waisman, H.A., & Harlow, H.F.   Abnormal social
behavior in phenylketonuric monkeys.   Journal of Abnormal
Psychology, 1970, 76(1), 62-68.

     Four rhesus monkeys, given a diet high in phenylalanine
     early in life, were compared with two control groups
     in learning and social behavior when all Ss were on a
     normal diet.   In comparison with the controls, the
     phenylketonuric (PKU) Ss were slow in learning a condi-
     tioned shock-avoidance task and showed extreme subnormal
     and inadequate social behavior.   This gross incompetence
     in social interaction was reflected in the inconsistence
     of dominance rankings in a competitive food situation
     and in the excessive hostility, excessive fear, and
     deficient play responses, both in the relatively unfamiliar
     playroom situation with familiar peers and in the home
     cage with unfamiliar stimulus monkeys.   These PKU Ss
     were normal in more primitive, more reflexive behaviors,
     e.g., behaviors reflecting activity, simple social,
     environmental, and self-stimulating behaviors.

88.   Chappell, P.F.   The dynamic proximal relationship of rhesus
macaque (Macaca mulatta) mother-infant pairs as affected by
temporary installation of physical barriers.   Ph.D. Thesis, George
Peabody College for Teachers, Nashville, Tennessee, 1972.

     Four male rhesus infants were reared for four months with
     their mothers in individual home cages attached to a
     large common play area.   Daily observations of the
     behaviors of these pairs provided data which showed that
     beginning at the birth of the infants an immediate
     synchronous responding existed between the pair members;
     with the elaboration of the infant's behavior, a simul-
     taneous elaboration of the mother's behavior occurred.
     Multivariate analyses showed Maternal Response Posture
     the best single predictor of infant behavior (R^2 = .35),
     whereas a composite set of predictors, Maternal Response
     Posture, Manipulation, and Vocalization, accounted for
     45% of the variance in a single index of infant behavior,
     distance from mother.   Installation of physical barriers
     for three days during four consecutive weeks after the
     infants were two months old limited communication in
     this already semi-restricted social environment and
     changed the reciprocal response patterns between the
     members of the individual pairs.   Primarily, the changes
     occurred in infant behavior, though concomitant changes
     also occurred in maternal behavior--but only when examined

in relationship to that of the infant. In general, the
barriers acted to constrain the behavior of the infants
and to increase the proximal relationship between the
pair members.

89.   Chappell, P.F., & Meier G.W.   Behavior modification in a mother-
infant dyad. <u>Developmental Psychobiology</u>, 1974, <u>7</u>, 296.

This note adds to the growing literature on the modifiability
of the behavior of isolate-reared primates and concerns the
mother-infant relations of an isolate-reared female rhesus
monkey (<u>Macaca mulatta</u>) maintained as a surgical control
in a follow-up study of neonatal brain damage (bilateral
electrolytic destruction of the inferior colliculi).
The mother was laboratory born and isolate reared; from
18 to 60 months she was part of a mixed group of 9 or
more age-peers from the study; from 60 months she remained
in a separate cage. These observations were made when
she was about 8 yr. old. The observations reveal the
potential for adaptation within the mother-infant dyad
and the reciprocating influence that each pair member
has on the other; they augment, in kind, those on infant-
infant pairs.

90.   Chappell. P.F., & Meier, G.W.   Modification of the response to
separation in the infant rhesus macaque through manipulation of the
environment. <u>Biological Psychiatry</u>, 1975, <u>10</u>(6), 643-657.

Rhesus mother-infant pairs were housed in a playpen
apparatus beginning just before the birth of four male
infants. The infants were separated from their mothers
four times beginning at a mean age of 218 days. In Type A
separations (I and IV) the infants were removed and
housed away from their familiar environment in a protected
setting; in Type B separations (II and III) the infants
remained in the familiar setting and mothers were removed.
One pair was separated every 2 weeks for 6 days; for a
particular infant, a mean of 8 weeks intervened between
each of the separations. On the basis of infant behavior
during separation, Type B separations appeared to have
a more deleterious effect on the infant: infants did
not show the typical behavioral signs of depression
under Type A housing conditions, whereas, under Type B
conditions, infants expressed the typical depressive
reaction to separation. However, comparisons of pre-

and postseparation behaviors in the mother-infant pairs
indicated that Type A separations were more perturbing.
Increases in ventral-ventral contact between mothers
and infants were greater following Type A separations
and increases in time at nipple occurred only after
Type A separations; infant grooming by mother increased
only after the first, a Type A, separation. Type B
separations may have affected mothers more severely in
that reciprocity between maternal cradling and infant
clinging was greater following Type B separations than
following Type A separations when infants clung
significantly more often than mothers cradled.

91.   Chow, K.L., Riesen, A.H., & Newell F.W.   Degeneration of
retinal ganglion cells in infant chimpanzees reared in darkness.
Journal of Comparative Neurology, 1957, 107, 27-42.

The present study reports the histological findings on the
retinas of three infant chimpanzees reared under various
conditions of light deprivation. All the retinas were
serially sectioned and every 10th section stained with
hematoxylin and eosin. Additional selected sections
were prepared with toluidine blue and protargol silver
stain. The retinas of two subjects show a disappearance
of ganglion cells, but intact receptive and bipolar cell
layers. One of the two animals was reared in the dark
from birth to 33 months old (with a few minutes of light
exposure daily) and the other was kept in total darkness
from 8 months to 24 months old. The retina of the third
chimpanzee who was reared in the dark from birth to 7
months old, but with one and a half hours of diffuse
light stimulation daily, is entirely normal. Behaviorly,
all three subjects showed various degrees of retardation
in visual development and visual discrimination learning.
The possibility that atrophy occurred in the retinal
ganglion cells of the first two animals as a result of
absence of function was discussed.

92.   Clark, D.L.   Immediate and delayed effects of early, intermediate, and late social isolation in the rhesus monkey.   Unpublished
Ph. D. dissertation, University of Wisconsin, 1968.

Sixteen infant rhesus monkeys were divided into 4 sex-
balanced groups of 4 animals each. Monkeys in the contact
group were provided within-group social experience between

1 and 15 months of age. Animals in 3 isolate groups
were isolated from social interaction for the first
6-months of life (early isolates), between 3 and 9-months
of age (intermediate isolates), and for the last 6-months
of the first year of life (late isolates). The data
suggest that physical social interaction experience is
a crucial factor in normal social development in the
rhesus monkey, and that early social experience prevented
the development of maladaptive behavioral tendencies.
The effect of social experience prior to 3-months of
age was the inhibition of the development of abnormal
crouching and rocking behavior, while the effect of social
experience prior to 6-months of age was the inhibition of
both crouching-rocking and abnormal or increased emotionality and assertiveness.

93.   Combs, C.M., Jacobson, H.N., McCroskey, D.L., Saxon, S.V.,
Stiehl, W., & Windle, W.F. A condition resembling "cerebral palsy"
in young Macaca mulatta surviving asphyxia neonatorum. Anatomy
Record, 1959, 133, 462.

Monkeys were delivered by cesarean section near term and
asphyxiated by keeping them inside intact amniotic sacs
until severe respiratory distress was evident. The
sacs were then opened. The pharynx was aspirated and
resuscitation accomplished with the use of an endotracheal
catheter. The motion picture will show one experiment
in which the time between the separation of the fetus
from the maternal circulation and the reestablishment
of oxygenation was 11 min. 58 sec. The first spontaneous
isolated gasps were seen 24 minutes following onset of
asphyxiation. By 38 minutes there was almost continuous
synchronous respiratory activity involving the neck muscles,
thorax, and diaphragm. The infant was in critical
condition for the first 4 days and required gavage
feeding up to day 6. It has continued to require hand
feeding for 67 days. Since the first week, this infant
has clearly and increasingly shown many of the classical
signs of cerebral palsy. It has had marked asthenia
from the beginning; it has ataxia and a steppage gait
in the hind limbs, more marked on the right side.
Choreiform movements of the head have been observed from
time to time. This infant has shown defects in learning
ability in that by day 67 it has not learned to feed
itself, a feat the normal infant monkey can do by 12
days.

94.   Crawford, M.P.   Dominance and the behavior of pairs of female
chimpanzees when they meet after varying intervals of separation.
Journal of Comparative Psychology, 1942, 33, 259-265.

      Fifty-four pairs of female chimpanzees, which had been
      separated from each other for periods varying from five
      weeks to three years, were observed for the first ten
      to thirty minutes after they were introduced into the
      same living cage. A running account of tne behavior of
      each animal was made, as it was observed, and later certain
      behavior items were transferred to a check sheet. Quan-
      titative and qualitative data were analyzed in terms of
      their value in predicting the results of food-dominance
      tests given to each pair shortly after the above observa-
      tions were completed. The figures show that the animal
      which, (1) enters the other animal's cage, (2) is groomed
      first by the other animal, and (3) attacks or bluffs
      the other animal is likely to prove dominant in the tests.
      Subordination is most reliably predicted from (1) retreating,
      (2) vocalization, and (3) grooming the other animal first.
      Great individual differences appeared between pairs of
      animals. In general, the intensity and complexity of
      behavior decreased as the same two animals were paired
      on successive occasions. Sexual behavior was relatively
      infrequent, and did not offer a consistent predictive
      index of subsequent dominance-test results.

95.   Cross, H.A. & Harlow, H.F.   Observation of infant monkeys by
female monkeys. Perceptual and Motor Skills, 1963, 16, 11-15.

      Interest in viewing infant monkeys was tested in nulliparous
      adult female monkeys, multiparous females delivered two
      to four months before the start of the experiment, and
      in gravid monkeys before and after delivery. The nulli-
      parous females and the gravid females before delivery
      showed no preference for baby-viewing, but mothers newly
      delivered or delivered some months earlier tended to
      view infants more frequently and for longer periods
      than the other groups.

96.   Cross, H.A., & Harlow, H.F.   Prolonged and progressive effects
of partial isolation on the behavior of macaque monkeys. Journal
of Experimental Research In Personality, 1965, 1, 39-49.

      The effects of partial social isolation were studied in

81 laboratory-born rhesus monkeys ranging in age from
1 to 7 years. Partial social isolation was achieved by
raising the monkeys from birth onward in individual wire
cages where they could see and hear other monkeys but
could not make physical contact with them. Partial social
isolation produced an early, relatively transient effect
on sucking orality and a more prolonged effect on chewing
orality. Overall disturbance decreased with age, but
threat responses, both externally and self-directed,
increased for a prolonged period of time. By and large
it appeared that there were progressive effects of partial
social isolation which exaggerated non-nutritional orality
and external and self-directed threat; there were clear-
cut differential developmental courses between the male
and female in terms of fear and aggression responses.

97. Cummins, M.S., & Suomi, S.J. Long-term effects of social
rehabilitation in rhesus monkeys. Primates, 1976, 17(1), 43-51.

A group of 19 rhesus monkeys was assembled and observed
in a large outdoor pen. Five experimental groups were
present: 3-year-old mother-peer-reared monkeys, surrogate-
rehabilitated isolates, socially-rehabilitated isolates,
surrogate-peers (who had served as "therapists" for the
preceeding group), and 1-year-old mother-peers. The
major results indicated first, that the surrogate-
rehabilitated isolates played socially significantly less
than both mother-peer groups, and second, that the
group as a whole seemed to display an abnormally low
overall level of behavior. Possible interpretations
and theoretical implications are discussed.

98. Davenport, R.K. Jr., & Menzel, E.W. Jr. Stereotyped behavior
of the infant chimpanzee. Archives of General Psychiatry, 1963,
8, 99-104.

The present paper is concerned with stereotypies in
infant chimpanzees. Evidence is presented to show that
stereotypy is related to rearing variables, developmental
status, immediate stimulus situation, and various forms
of ongoing activity. Most strikingly, stereotypies are
phenomena unique to infants raised in restricted
environments. They commence within the first few months
of life and persist into adulthood. These behaviors
show marked resemblances to behaviors of human beings with
certain pathological conditions.

99. Davenport, R.K. Jr., & Rogers, C.M. Intellectual performance
of differentially reared chimpanzees: I. Delayed response. American
Journal of Mental Deficiency, 1968, 72, 674-680.

    Seven chimpanzees separated from their mothers at birth
    and raised for two years in a restricted environment were
    compared on spatial delayed response tasks to eight wild-
    born enriched environment chimpanzees when both groups
    were between 7 and 9 years of age. Restricted Ss were
    initially inferior but with experience closely approached
    the wild-born performance level. Differences in performance
    between the groups and delays and improvement with
    experience are explained in terms of relative differences
    and changes in task-oriented and non-task-oriented
    behaviors. Findings are related to learning problems and
    IQ test performance of human children reared in unstim-
    ulating environments.

100. Davenport, R.K. & Rogers, C.J. Differential rearing of the
chimpanzee: A project survey. In G.H. Bourne, (Ed.), The Chimpanzee.
Vol. 3, 1970, Basel, Switzerland: S. Karger, 337-360.

    Contrary to results reported for mother-deprived and
    environmentally restricted human infants, the restricted-
    environment chimpanzees weighed more and were in better
    physical health than mother-reared infants. In addition,
    they did not develop the adverse physical effects or
    apathy reported in human infants. Infant chimpanzees
    separated from the mother very early in life invariably
    develop repetitive stereotyped behaviors. Restricted
    rearing produces a deep and pervasive fearfulness and
    timidity. However, this decreases markedly with time
    and experience. Individual differences among restricted
    environment animals (even those raised in the same
    condition) are at least as great as individual differences
    among wildborn animals (in the laboratory) of the same
    age. The restricted group was deficient in learning and
    adaptability when compared to wildborn subjects. It
    appears that chimpanzees are less severely affected by
    social deprivation than are monkeys, and partial recovery
    from abberant behavior does occur in the chimpanzee, in
    contrast to what has been reported in the monkey.

101. Davenport, R.K. Jr., Menzel, E.W., & Rogers, C.M. Maternal
care during infancy: Its effect on weight gain and mortality in
the chimpanzee. American Journal of Orthopsychiatry, 1961, 31(4),
803-809.

    The present paper considers the effects of three conditions
    of rearing on weight gain and mortality. Weight and
    mortality statistics for three groups of infant chimpanzees
    differing grossly in the degree and kind of maternal care
    during infancy have been provided by a series of studies
    at Yerkes Laboratories. The infants were reared in one
    of three conditions: 1) with the mother, 2) in the
    experimental nursery, or 3) in an extremely restricted
    and impoverished environment. A comparison of weight
    and mortality in these three groups permits a direct
    assessment not only of maternal deprivation but also of
    virtually complete absence of maternal care.

102. Davenport, R.K. Jr., Menzel, E.W. Jr., & Rogers, C.M.
Effects of severe isolation on "normal" juvenile chimpanzees.
Archives of General Psychiatry, 1966, 14, 134-138.

    Previous studies have demonstrated that (A) Infant
    chimpanzees raised from birth in conditions of extreme
    restriction remain in better health and weigh more than
    mother-reared animals (contrary to findings on human
    infants). (B) Infant chimpanzees deprived of the mother
    very early develop odd and persistent motor patterns,
    stereotypies. In the present experiment three wild-
    born, and for several months mother-reared, infants were
    placed in the condition of extreme restriction when
    approximately 18 months of age. During six months of
    isolation all three remained in good health, consistently
    gained weight, and did not develop the stereotyped
    patterns previously noted. These results are compared
    to the findings reported in similar but less drastic
    experiments with human infants.

103. Davenport, Jr., R.K., Rogers, C.J., & Menzel, Jr., E.W.
Intellectual performance of differentially reared chimpanzees: II.
Discrimination learning set. American Journal of Mental Deficiency,
1969, 73(6), 963-969.

Eight near-adult chimpanzees that had been reared for the
first 2 years of life in restricted laboratory environments
were inferior to 8 wildborn controls both in pretraining
and in discrimination learning set formation. Since both
groups had shared the same cages and test experiences
after the age of 2-3 years, the role of early experience
is emphasized. The deficit of restricted Ss seemed to
entail an inability to inhibit old responses (e.g.,
they made significantly more "stimulus perseveration errors"
than controls) and a tendency to be easily distracted
into "irrelevant" behaviors by "irrelevant" stimuli.

104. Davenport, R.K. Jr., Rogers, C.M., & Rumbaugh, D.M. Long-term
cognitive deficits in chimpanzees associated with early impoverished
rearing. Developmental Psychology, 1973, 9(3), 343-347.

Six adult chimpanzees that had been reared for the first
two years of life in restricted laboratory environments
were inferior in cognitive skills to eight wild-born
control subjects, as assessed by Transfer Index testing.
Since both groups had shared the same cages and test
experiences after three to four years of age, the role
of early experience in cognitive development was under-
scored.

105. Deets, A.C, & Harlow, H.F. Adoption of single and multiple
infants by rhesus monkeys mothers. Primates, 1974, 15(2-3), 193-204.

Adoption was studied in Macaca mulatta. Multiparous mothers
were separated from their biological offspring within
hours following parturition, and 2½ days later, on the
average, they were offered neonates for adoption. These
foster infants had been separated from their biological
mothers shortly after birth and averaged 2 3/4 days old
when subjected to adoption. Mothers that were offered
a single neonate immediately and completely adopted the
foster infant. But mothers that were offered two infants
at the same time were ambivalent. Many infants were
rejected initially, although with one partial exception,
the mothers eventually accepted and cared for two infants.
It was concluded that the macaque maternal affectional
system may be biased toward accepting and nurturing one
infant at a time.

106.   Deets, A.C., Harlow, H.F., Singh, S.D., & Blomquist, A.J.
Effects of bilateral lesions of the frontal granular cortex on the
social behavior of rhesus monkeys. Journal of Comparative and
Physiological Psychology, 1970, 72, 452-461.

   Eleven monkeys with bilateral lesions of the frontal granular
   cortex and 11 matched controls were observed during
   pairings with each of 12 stimulus monkeys. The frontal
   subjects were more withdrawn and distressed than the
   control animals. The operated animals showed less
   proximity and contact with the stimulus animals, directed
   less exploration toward the inanimate environment, and
   displayed more fear grimacing, screeching, and other
   disturbance behaviors. The stimulus animals, in turn,
   interacted less with the frontal monkeys than with the
   controls. They spent less time near the operated animals,
   and less frequently mounted and presented to them. Overall,
   the operated subjects directed more challenges toward the
   male stimulus animals, threatening them more than was
   appropriate for the circumstances, although levels of
   overt physical aggression were no longer, and in some
   cases depressed, in the monkeys with frontal lesions.

107.   Deutsch, J., & Larsson, K.   Model-oriented sexual behavior in
surrogate reared rhesus monkeys.   Brain, Behavior and Evolution,
1974, 9, 157-164.

   Surrogate- and mother-reared rhesus monkeys were exposed
   to a stationary cloth-covered model. All four of the
   surrogate-reared animals displayed social and sexual
   patterns toward the object while none of the six mother-
   reared animals did so. While both of the surrogate-
   reared males performed repeated mounts on the object,
   one of these males was capable of consistently performing
   complete sequences of properly oriented mounts interspersed
   with grooming and terminating in ejaculation. When
   tested with female monkeys, neither of the surrogate-
   reared males performed repeated mounts. The establishment
   and maintenance of contact requisite for sexual interaction with the females appeared to be prevented by the
   surrogate-reared males' tendencies to either passivity
   and withdrawal or erratic aggressiveness. Since the
   presence of a reciprocally active female monkey disrupted
   the expression of sexual patterns elicited by the unrespon-
   sive model, it was suggested that early adjustment to a
   noninteractive 'mother' interferes primarily with the
   surrogate-reared monkey's perception of the social object
   and ability to organize interactive social events.

108. Draper, W.A., & Bernstein, I.S. Stereotyped behavior and cage size. Perceptual and Motor Skills, 1963, 16, 231-234.

Twelve feral adolescent rhesus monkeys were observed
individually during ten 5-min. periods in each of three
different-sized cages. Stereotyped and cage-oriented behavior
occurred most frequently in the small cage, sometimes in the
medium cage, and never in the large cage. It was concluded
that spatial restriction which does not permit "normal"
locomotor behavior, e.g., running, climbing, etc., results
in substitute motor expression which frequently takes
the form of repetitive stereotyped movements.

109. Drickamer, L.C., & Vessey, S.H. Group changing in free-
ranging male rhesus monkeys. Primates, 1973, 14(4), 359-368.

Group changing behavior of male Macaca mulatta was studied
over a six-year period at the rhesus monkey colony on
two coastal islands at La Parguera, Puerto Rico. Males
first left their natal group at a mean age of 47 months
and became solitary for the first time at a mean age
of 64 months; all had left their natal groups by seven
years of age. Age, mating season, sex ratios of adult
males and females in the social bands, and geographical
barriers all had significant effects on the group shifting.
Population size, rank of mother or being an orphan did
not significantly affect the changing process. Two
factors, age (size) and seniority in the group, were
important in determining a male's rank in his new group.

110. Early, K. Father love. The Sciences, 1974, 14(8), 24-25.

This article discusses the work of Mitchell et al. on
the care of infants by male rhesus monkeys. In a number
of experiments, 9 to 11 year-old males were paired with
an adoptive infant cage-mate for several months during
the formative socializing period of the species, and
their behavior compared to female adults and their natural
infants. The results suggest that rhesus monkeys have
complementary parenting systems with sex differentiated
components.

111. Eastman, R.F., & Mason, W.A. Looking behavior in monkeys
raised with mobile and stationary artificial mothers. Developmental
Psychobiology, 1975, 8(3), 213-221.

    Rhesus monkeys raised with mechanically driven mobile
    artificial mothers for the first 10 months of life looked
    more at other animals and were more sensitive to stimulus
    differences than monkeys raised on stationary surrogates.
    Both groups looked less and were not as discriminating
    as wild-born monkeys. The results are consistent with
    other evidence suggesting that one of the long-range
    consequences of maternal mobility is to reduce emotional
    responsiveness to novel situations.

112. Elias, M.F., & Samonds, K.W. Exploratory behavior of cebus
monkeys after having been reared in partial isolation. Child
Development, 1973, 44, 218-220.

    Six cebus monkeys were reared until the age of 6 months in
    one of two conditions: with or without exposure to other
    monkeys and a variety of objects in a gang cage. Develop-
    ment was observed weekly, and exploratory behavior was
    tested at 6 months of age. The isolated monkeys were
    slightly retarded in development and displayed more stereo-
    typed behavior. They explored less in a novel environment
    and appeared to be more timid. The findings are consistent
    with behavior observed in rhesus monkeys reared under
    similar conditions, thereby broadening the base from
    which to extrapolate to human development.

113. Elias, M.F., & Samonds, K.W. Exploratory behavior and activity
of infant monkeys during nutritional and rearing restriction.
American Journal of Clinical Nutrition, 1974, 27, 458-463.

    Diminished exploration and activity were found in cebus
    monkeys during a period of protein or calorie restriction
    or isolated rearing imposed from 8 to 26 weeks of age.
    Calorie restriction reduced activity somewhat. Rearing
    affected exploratory behavior more than did diet, but
    double restriction in protein and rearing produced
    particularly severe impairment.

114. Elias M.F., & Samonds, K.W. Protein and calorie malnutrition
in infant cebus monkeys: growth and behavioral development during
deprivation and rehabilitation. American Journal of Clinical
Nutrition, 1977, 30, 355-366.

    The growth and development of 32 cebus monkeys were studied
    during a period of insult in nutritional or rearing condi-
    tions and after rehabilitation. Eight experimental groups
    of four animals each were subjected to one of four diets--
    control, protein restricted, calorie restricted, and protein-
    calorie restricted, and one of two rearing conditions--
    partial isolation or a comparatively enriched condition--in
    a 4 x 2 factorial design. The period of insult from 2
    to 6 months of age was followed by 6 months of rehabilita-
    tion in both diet and rearing conditions. It was found
    that only diet affected physical growth, but both diet
    and rearing affected behavioral development and exploratory
    behavior. Whereas calorie deficiency produced a direct
    effect on behavior independent of rearing conditions,
    protein deficiency produced an effect only in combination
    with rearing restriction. The effect of protein-calorie
    deficiency had some characteristics in common with each
    of the other deficiencies. Retardation in rate of
    behavioral development was less severe than retardation
    in growth, most notably in the protein-restricted,
    enriched-rearing group, producing animals who were
    behaviorally mature for their size. All groups caught
    up in physical growth during rehabilitation but the
    protein-calorie restricted group failed to recuperate
    completely in exploratory behavior.

115. Erwin, J. Responses of rhesus monkeys to separation
vocalizations of a conspecific infant. Perceptual and Motor Skills,
1974, 39, 179-185.

    The vocalizations of a rhesus monkey infant were recorded
    immediately following separation of the infant from its
    mother. The recorded vocalizations were played to 12
    young adult rhesus monkeys, 6 of each sex. The adults
    were less active during the time when the recorded
    vocalizations were being played than during equivalent
    periods of time before and after the stimulus tape
    had been played. Ss also spent more time near the
    stimulus source and looked at it longer when infant
    vocalizations were audible. Two of the females responded
    to the infant sounds immediately by looking down at their
    breasts and by displaying retrieval or support postures
    typical of mother monkeys with small infants.

116. Erwin, J.   The development and persistence of rhesus macaque
(Macaca mulatta) peer attachments (PhD. dissertation, University of
California Davis, 1974), Dissertation Abstracts International, 1975,
35B, 4214.

    Studies of free-ranging rhesus macaques (Macaca mulatta)
    have suggested that long-term emotional attachments between
    individual group members are partly responsible for the
    cohesive group structures which occur in this species.
    A series of laboratory studies was designed to explore
    the nature of emotional ties between juvenile and adolescent
    rhesus peers of both sexes.

117. Erwin, J., & Deni, R.   Strangers in a strange land: Abnormal
behaviors or abnormal environments?   In J. Erwin, T. Maple, and
G. Mitchell (Eds.), Captivity and behavior.   New York: Van Nostrand
Reinhold, 1979, 1-28.

    This chapter reviews some of the information on abnormal
    behavior of primates and the environmental factors asso-
    ciated with them.   Some definitions of abnormal behavior
    are discussed as well as a number of environmental
    factors that may influence behavior.   Examples of the
    role of environmental influences in the development or
    prevention of maladaptive behavior among captive primates
    are cited.

118. Erwin, J., & Flett, M.   Responses of rhesus monkeys to reunion
after long-term separation: Cross-sexed pairings. Psychological
Reports, 1974, 35, 171-174.

    Twelve 5½-year-old rhesus monkeys, six of each sex, were re-
    united with other-sexed peers with which they had been paired
    for 6 months during early adolescence. Responses of Ss
    to reunion with familiar animals were compared with their
    responses to pairing with unfamiliar other sexed peers.
    The members of familiar pairs had been separated for
    nearly 2 yr. prior to the reunion described here.
    Familiar animals never aggressed one another, but males
    aggressed unfamiliar females. Ss directed more threats
    toward unfamiliar than familiar animals. Males mounted
    familiar females more often than unfamiliar ones, but
    it was apparent that heterosexual attractiveness in
    general was not based exclusively on familiarity.

119. Erwin, J., & Mitchell, G. Initial heterosexual behavior of
adolescent rhesus monkeys (Macaca mulatta). Archives of Sexual
Behavior, 1975, 4(1), 97-104.

       This report describes the first heterosexual encounters
       of twelve 3-year-old rhesus monkeys, six of each sex.
       The subjects were all laboratory-born, and were reared
       in wire cages for the first 8 months of life, accompanied
       only by their mothers. After these animals were weaned,
       each was placed in a cage with another animal of the same
       age and sex. The pairs formed in this way remained
       intact until the animals were 2 years old. When hetero-
       sexual dyads were formed, some of the subjects attacked
       and bit the animal with which they had been paired,
       while the members of other pairs established almost
       immediate rapport as evidenced by mutual grooming and
       adequate sexual behavior. The initial sexual interactions
       of most pairs were uncoordinated, but all eventually
       demonstrated qualitatively species-typical patterns of
       sexual behavior. Despite their immaturity, two of these
       pairs succeeded in producing offspring, both of which
       were healthy.

120. Erwin, J., Brandt, E.M., & Mitchell, G. Attachment formation
and separation in heterosexually naive preadult rhesus monkeys
(Macaca mulatta). Developmental Psychobiology., 1973, 6(6),
531-538.

       Six heterosexual pairs of 3-yr-old rhesus monkeys which
       had no previous experience with other-sexed peers were
       allowed 1, 2, or 3 weeks to form social attachments.
       Attachment strength was evaluated by observation of the
       subjects' responses to separation and reunion. Those
       pairs which were together for the shortest period of
       time exhibited the greatest amount of disturbance at
       separation, but they were also the most disturbed prior
       to separation and at reunion. The behavior of females
       was influenced more by length of pairing than was the
       behavior of males.

121. Erwin, J., Maple, T., & Welles, J.F. Responses of rhesus
monkeys to reunion: Evidence for exclusive and persistent bonds
between peers. Contemparary Primatolgy 5th International Congress
of Primatology, 1974, 254-262 (Karger/Basel)

Responses of rehsus monkeys to reunion with like- and other-
sexed peers with which they had lived during juvenility
or adolescence were compared with their responses to
pairing with strangers. Familiar animals had been
separated from each other for about 2 years prior to
reunion. Subjects displayed less aggression and disturbance and more affiliation with familiar than with
unfamiliar stimulus animals.

122. Erwin, J., & Mitchell, G., & Maple, T. Abnormal behavior in
non-isolate-reared rhesus monkeys. Psychological Reports, 1973,
33, 515-523.

The research reported here documents the existence of
self-directed aggression in non-isolate-reared rhesus
monkeys. Ss for this series of studies were reared with
their mothers for most of the first year of life, and
each animal experienced constant social access to a
like-sexed peer throughout its second year of life.
Significant amounts of social contact were also provided
during the third and fourth years. Ss were observed
in social situations at approximately 1, 2, 2½, 3, and
4½ years of age, and some self-biting was observed at
each age. Self-aggression occurred primarily in semi-
stressful contexts which apparently did not allow appropriate outward-directed expression of emotion. The use of
socially reared animals as models for the study of self-
directed aggression is suggested as an alternative to
the use of isolate-reared Ss.

123. Erwin, J., Mobaldi, J., & Mitchell, G. Separation of rhesus
monkey juveniles of the same sex. Journal of Abnormal Psychology,
1971, 78(2), 134-139.

The nature of mother-infant attachments in monkeys has
been studied by disrupting the attachment once it has
been established. Juvenile-juvenile separations were
studied here, and a Bowlby separation syndrome was
observed and compared to mother-infant separations
reported elsewhere. Similarities were greatest during
the period of separation itself. The differences between
the two dyads were most marked during reunion. Juvenile
monkeys initially ignore or avoid each other during
reunion while mothers and infants usually directly return
to one another.

124. Erwin, J., Maple, T., Mitchell, G., & Willott, J. Follow-up
study of isolation-reared and mother-reared rhesus monkeys paired
with preadolescent conspecifics in late infancy: Cross-sex pairings.
Developmental Psychology, 1974, 10(6), 808-814.

    Eight rhesus monkeys, four of each sex, which had been
    reared either with their mothers or in social isolation
    during early infancy, were paired with preadolescent
    conspecifics during late infancy. Some isolate-reared
    subjects appeared to have gained from this social experience.
    At nearly 3 years of age, each subject was paired
    (sequentially) with two other-sexed animals (one isolate-
    and one mother-reared) after having been housed alone for
    the entire second (and most of the third) year of life.
    Little social interaction occurred between the members
    of any pair. Several significant effects of rearing
    experience indicated that social experience with pre-
    adolescents in late infancy did not permanently reverse
    the deleterious effects of early isolation. However, the
    mother-reared subjects also failed to establish social
    rapport with one another. Comparison of these results
    with those of similar research employing mother-reared
    animals which had received supplemental social experience
    during the second year of life suggests that such social
    experience is of some importance for optimal development
    of later sociosexual behavior.

125. Erwin, J., Maple, T., Willott, J. & Mitchell, G. Persistent
peer attachments of rhesus monkeys: Responses to reunion after
two years of separation. Psychological Reports, 1974, 34,
1179-1183.

    Twelve 4½-year-old rhesus monkeys, 6 of each sex, were reunited
    with like-sexed peers with which they had spent their
    second year of life. Responses of Ss to reunion with
    familiar animals were compared with their responses
    to pairing with unfamiliar like-sexed peers. The members
    of familiar pairs had been separated for more than 2
    years prior to reunion. Ss displayed less aggression,
    fear/submission, and disturbance, and more affiliation,
    while paired with familiar peers than with unfamiliar
    peers.

126. Evans, C.S. Methods of rearing and social interaction in
Macaca nemestrina. Animal Behavior, 1967, 15, 263-266.

   The social interactions of two similar groups, each of
   four infant pig-tailed macaques, were recorded under
   standard conditions and compared for effects of their
   early experience. One group, mother-reared and allowed
   to live in a relatively unrestricted social environment,
   served as controls. The other group comprised infants
   removed from their mothers at birth and reared under
   restricted social and physical conditions. Play and
   communicative behaviours were significantly less frequent
   in the group of socially-restricted infants compared
   with experienced infants. Only in the isolated and
   socially inexperienced group did fear-grimacing, aggression digit-sucking and self-clasping appear, effects
   paralleling those also attributed to maternal absence
   and insufficient contact with peers in rhesus infants.
   The results indicate that early social deprivation
   interferes with the development of social responsiveness
   in the pigtailed macaque as in the rhesus.

127. Fittinghoff, N.A., Jr., Lindburg, D.G., & Mitchell, G.
Failure to find polydipsia in isolation-reared monkeys.
Psychonomic Science, 1971, 22(5), 277-278.

   The water consumption of six adult male monkeys, reared
   in social isolation, was compared with that of five feral-
   reared controls. Isolate polydipsia, reported elsewhere,
   was not found in the present study.

128.Fittinghoff, N.A., Jr., Lindburg, D.G., Gomber, J., & Mitchell, G.
Consistency and variability in the behavior of mature, isolation-
reared, male rhesus macaques. Primates, 1974, 15(2-3), 111-139.

   Self-punishments, hyperaggressiveness, stereotyped behaviors, bizarre movements, and masturbation are more
   typical of adult-male isolates than of adult-male wild-
   born rhesus. These differences persist into the 13th
   year. Social exploration and cage-shaking are also
   depressed in isolates. A number of speculative explanations are offered for many isolate behaviors. An
   individual animal's abnormalities not only change but
   decrease with age and the kinds and frequencies of
   abnormalities decrease as isolates habituate to a new

situation.   The behavior of controls is much less variable
than is the behavior of isolates.   Saluting and eyeball
pressing correlate with reduced levels of arousal.
Isolate "abnormalities" are viewed as "normal" responses
to an altered ecology.

129.   Fitz-Gerald, F.L.   Effects of D-amphetamine upon behavior of
young chimpanzees reared under different conditions.   Proceedings
of the Fifth International Congress of the Colloquium Internationale
Neuropsychopharmacoloquieum, Excerpta Medica International Congress
Series No. 129, 1966, 1226-1227.

    Nine restriction-reared and nine mother-reared juvenile
    chimpanzees were tested in a series of experiments designed
    to study the effects of arousal level and opportunity
    for alternative activities upon various patterns of
    behavior which are differentially characteristic of
    primates reared with or without their mothers.

130.   Foley, J.P. Jr.   First year development of a rhesus monkey
(Macaca mulatta) reared in isolation.   Journal of Genetic Psychology,
1934, 45, 39-103.

    The subject of this investigation was a male rhesus monkey
    separated from his mother at the age of three days and
    reared in social isolation for a period of one year.
    The physical development, sensorimotor and simple behavioral
    development, and complex behavioral development of the
    subject during this period are discussed.

131.   Foley, J.P. Jr.   Second year development of a rhesus monkey
(Macaca mulatta) reared in isolation during the first 18 months .
Journal of Genetic Psychology, 1935, 47, 73-97.

    This paper reports data on the 2nd year development of
    a rhesus monkey (M. mulatta) reared for one and a half
    years in isolation from its mother and other members
    of the species, and subsequently housed with other monkeys
    of a similar and different species.   Developmental
    aspects of early behavior and their implication for
    further research are discussed.

132. Fox, M.W. Psychopathology in man and lower animals. Journal of the American Veterinary Medical Association, 1971, 159(1), 66-77.

This article reviews several behavioral disorders in lower animals which have been well-recognized in man. Human traits, symtoms, and specific syndromes associated with general behavior disturbances are discussed with reference to comparable disorders in lower animals.

133. Fox, S.S. Self-maintained sensory input and sensory deprivation in monkeys: A behavioral and neuropharmacological study. Journal of Comparative and Physiological Psychology, 1962, 55, 438-444.

The present series of experiments was designed to investigate behavior by which sensory input levels are maintained by a monkey's bar pressing for light. Three experiments were conducted with monkeys to determine the part played by three separate aspects of the behavior: deprivation, adaptation, and pharmacology. In Experiment 1, constant and regular rates of response were observed to occur when the response was bar pressing for 0.5 sec. of light. Rate increased with increasing sensory deprivation. In Experiment 2, by partially compensating with light during the test for the need for light produced by deprivation, rates of response were decreased for all deprivation durations. In Experiment 3, injected amphetamine increased the rate of response for light. It was interpreted that this drug's action indicated an increased requirement of the animal for sensory input. Additionally, studies were mentioned in which the same animals were implanted with deep electrodes and allowed to bar press for light.

134. Fragaszy, D.M. & Mitchell, G. Infant socialization in primates. Journal of Human Evolution, 1974, 3(6), 563-574.

This review considers evolutionary trends, functions, stages, and current issues in primate socialization. The concept of social "set" is discussed in the context of the interactional nature of socialization, where the infant is regarded as a primary effector in its own socialization. Laboratory approaches and measurement techniques are reviewed briefly.

135. Frank, R.G., Gluck, J.P., & Strongin, T.S. Response suppression
to a shock-predicting stimulus in differentially reared monkeys
(Macaca mulatta). Developmental Psychology, 1977, 13(3), 295-296.

  Rhesus monkeys reared in social isolation for the first
  9 months of life and monkeys reared with social contact
  were trained to lever press for food reinforcement on a
  variable interval 60 schedule of reinforcement as adults.
  A 3-minute auditory conditioned stimulus that terminated
  with the delivery of an electric foot shock was then
  presented at random intervals during each test session.
  Results indicated that though preshock response rates
  did not differ, the isolates failed to suppress responding
  in the presence of the shock-predicting conditioned
  stimulus at the same rate or to the same degree as did
  socially reared animals.

136. Furchner, C.S., & Harlow, H.F. Preference for various
surrogate surfaces among infant rhesus monkeys. Psychonomic
Science, 1969, 17(5), 279-280.

  Two infant monkeys were tested in terms of contact
  time for preference among four surrogate mothers with
  different body surfaces. Durations of contact differed
  significantly among the surrogate surfaces, which ranked
  from greatest to least time in the following order:
  cotton sock, rayon and vinyl (equal), and sandpaper.
  The results indicate that infant monkeys have clear-cut
  preferences among surrogate surfaces.

137. Gallup, G.G. Jr. Chimpanzees: Self-recognition. Science,
1970, 167, 86-87.

  After prolonged exposure to their reflected images in
  mirrors, chimpanzees marked with red dye showed evidence
  of being able to recognize their own reflections.
  Monkeys did not appear to have this capacity.

138. Gallup, G.G. Minds and mirrors. New Society, 1971, 18,
975-977.

  This article focuses on the capacity for self-recognition
  in chimpanzees as affected by early experience. Results

indicate that laboratory-born chimps raised in social
isolation, do not seem capable of learning to recognize
their reflections after extensive exposure to mirrors.
Following recovery from anesthesia, attention to marked
areas is virtually absent and typical self-recognition
patterns are nonexistent.

139. Gallup, G.G. Jr. Towards an operational definition of self
awareness.   In R.H. Tuttle (Ed.), Socioecology and psychology of
primates. The Hague: Mouton, 1975, 309-341.

The purpose of this paper is two-fold. First, an attempt
is made to review the literature on the psychological
properties of mirrors as determined by research on a
variety of species, and second to show how such surfaces
may provide a means by which an objective assessment
of self-awareness can be accomplished.

140. Gallup, G.G. Jr., & McClure, M.K. Preference for mirror-
image stimulation in differentially reared rhesus monkeys. Journal
of Comparative and Physiological Psychology, 1971, 75(3), 403-407.

When given a choice between viewing themselves in a
mirror or looking at another monkey, feral rhesus monkeys
seemed to prefer viewing the conspecific. On the other
hand, surrogate-reared animals spent appreciably more
time viewing the mirror and interacted more with their
reflection than with the conspecific. The results were
interpreted in terms of the effects of early social
isolation and the psychological properties of mirrors.

141. Gallup, G.G. Jr., McClure, M.K., Hill, S.D., & Bundy, R.A.
Capacity for self-recognition in differentially reared chimpanzees.
The Psychological Record, 1971, 21, 69-74.

By way of replicating previous work, wild-born chimpanzees
given prolonged exposure to mirrors learned to recognize
their own reflections. Chimpanzees born in captivity
and reared in social isolation did not, however, show
behaviors suggestive of self-recognition. The results
have possible implications for a theory which proposes
that self-concepts arise out of interpersonal
relationships.

142.   Gandolfo, R.L. , & Schlottman, R.S.   Social isolation of socially sophisticated Java (Macaca irus) monkeys.   Developmental Psychology, 1970, 2(1), 150.

This study presents behavioral data taken for two, 1-year-old, female Java infants housed in individual cages.   There were three experimental phases: control (2 weeks), isolation (8 weeks), and peer reunion (2 weeks). In contrast to studies isolating monkeys from birth, there was an increase in social interaction following isolation.   It appears that once social interaction activity becomes established, it is fairly resistant to the stress of a short-term isolation experience. Since these animals appeared to be clearly disturbed during isolation, long periods of isolation might generate abnormal social behavior.   The isolation exper- ience did not appear to produce overly aggressive behavior in these two subjects.

143.   Ganz, L., & Riesen, A.H.   Stimulus generalization to hue in the dark-reared macaque.   Journal of Comparative and Physiological Psychology, 1962, 55(1), 92-99.

A study was conducted to investigate whether the stimulus- generalization gradients to hue is present in the macaque, prior to experience with varied hues and in what manner stimulus experience might modify the gradient.   A Hue- Naive Group was reared in total darkness, and an Experienced Group was simularly reared except in a normally lit and visually patterned environment until age ten weeks.   Both groups were treated identically there- after.   Seated in a primate chair, the infants emitted a keypress response during the monochromatic stimulus, reinforced by sucrose solution(15-sec. VI), which was gradually discriminated with respect to Ss, a blackout interval (15-sec. delay of reinforcement).   Generali- zation was measured under extinction during seven testing days to a 180-mμ spectrum.   Orthogonal polynomial analysis revealed decremental gradients in both groups, steeper and concave upward in the Hue-Naive Group, relatively shallow and linear in the Experienced Group.

144. Gardner, E.L., & Gardner, E.B. Orientation of infant macaques
to facially distinct surrogate mothers. Developmental Psychology,
1970, 3(3), 409-410.

   The present study raises the possibility that infant
   monkeys may show orientation preferences toward one type
   of surrogate-mother face over another. Two infant
   rhesus monkeys (M. mulatta) 48 and 59 days old, were
   removed from their mothers and each placed in a cage
   containing two terrycloth covered mother surrogates
   like those used by Harlow. The surrogates were identical
   except for the faces. One of each surrogate pair had
   a square face similar to those on Harlow's wire mothers;
   the other had a round face similar to those on Harlow's
   cloth mothers. Infants showed clear preferences for the
   square face used on Harlow's wire surrogates. These
   results suggest the possibility that attentional variables
   of the sort already demonstrated in human infants may
   exist during certain developmental periods in infrahuman
   primates as well.

145. Gluck, J.P., & Harlow, H.F. The effects of deprived and
enriched rearing conditions on later learning: A review. In
L.E. Jarrard (Ed.), Cognitive processes of nonhuman primates.
New York: Academic Press, 1971.

   This chapter reviews the literature on two general types
   of early experience: enrichment or supernormal stimulation,
   and deprivation or subnormal stimulation.

146. Gluck, J.P., & Sackett, G.P. Frustration and self-aggression
in social isolate rhesus monkeys. Journal of Abnormal Psychology,
1974, 83(3), 331-334.

   The frequency of self-aggressive behavior emitted by
   partial social isolate and socialized rhesus monkeys
   was compared under baseline living conditions and a
   continuous reinforcement-extinction schedule of lever
   pressing. It was found that under baseline conditions,
   male isolates showed higher levels of self-aggressive
   behavior than did the female isolates. Further, frustration produced by extinction of the lever-pressing response
   produced a significant but transitory intensification
   of self-aggressive behavior. A theoretical interpretation of this relationship is briefly discussed.

147. Gluck, J.P., & Sackett, G.P. Extinction deficits in socially
isolated rhesus monkeys (Macaca mulatta). Developmental Psychology,
1976, 12(2), 173-174.

    Monkeys reared in total social isolation, partial isolation,
    and normally with access to mothers and peers were compared
    on the rate of acquisition of a simple operant response,
    maintenance of the response on a schedule of continuous
    reinforcement, and the rate of extinction. Criterion
    measures for acquistion of responding revealed that
    isolates took longest to reach the magazine training
    non-pause criterion, but took significantly less time
    to complete the continuous reinforcement criterion
    compared to either the wire cage or peer-raised groups.
    Isolates also took more time to emit the first lever
    press. Differences in operant behavior did appear
    during extinction, with isolates making more unrewarded
    responses and taking longer to stop responding in the
    face of nonreward.

148. Gluck, J.P., Harlow, H.F., & Schiltz, K.A. Differential
effect of early enrichment and deprivation on learning in the
rhesus monkey (Macaca mulatta). Journal of Comparative and
Physiological Psychology, 1973, 84(3), 598-604.

    Data are presented to both support and extend research
    findings concerning the effects of early experience on
    nonhuman primate learning ability. Enriched monkeys
    tested in their home living environment performed more
    proficiently than monkeys separated from their living
    environments and tested in an adjoining room. Further,
    monkeys reared in enriched environments were superior
    to partially isolated controls on the complex oddity
    tasks but not on 2-choice discrimination or delayed-
    response problems.

149. Gomber, J.M. Caging adult male isolation-reared rhesus
monkeys (Macaca mulatta) with infant conspecifics. Dissertation
Abstracts International, 1976, B36, 5302.

    In this study adult male isolation-reared rhesus monkeys
    were paired with 1-month-old conspecifics. Each isolate
    male-infant pair remained together for a 7-month rearing
    period, followed by a brief separation and reunion.
    Although isolate-male-reared infants and infants reared

with control males appeared to experience more stress
than mother-reared infants, no differences among groups
were great enough to warrant a judgment of abnormality
for the isolate-male-reared infants. In general, the
two groups of male-reared infants were most similar to
each other, but not greatly different from the mother-
reared infants. Direct comparisons between isolates
and control adult males also revealed that early isolation deficits are not as severe as reports have indicated.
Even though they exhibited less positive social behavior
overall, the isolate males increased over time in more
positive behavior items than did control males.

150. Gomber, J., & Mitchell, G. Preliminary report on adult male
isolation-reared rhesus monkeys caged with infants. Developmental
Psychology, 1974, 10(2), 298.

Isolation-reared monkeys have been described as excessively
fearful, hyperaggressive, lacking in communicative skills,
and adverse to physical contact. Prolonged interaction
was enforced by housing each fully mature, isolation-
reared rhesus male alone with a young infant for seven
months. The male's lack of hyperaggression, exhibition
of and response to social cues, frequency and form of
play grooming and physical contact, and adequacy as an
object of social attachment suggest the necessity to
reevaluate current ideas on the social capabilities
of isolation-reared monkeys.

151. Goy, R.W., & Phoenix, C.H. The effects of testosterone
propionate administered before birth on the development of behavior
in genetic female rhesus monkeys.   In C. Sawyer and R. Gorski
(Eds.), Steroid hormones and brain function. Berkeley: University
of Cal. Press, UCLA Forum in Medical Sciences, No. 15, Chapter 19,
1972.

This chapter discusses studies undertaken by the authors
which have been aimed at determining the influence of
early androgen on dimorphic psychological traits which
are independent of any hormonal activation. Rhesus
monkeys, reared under laboratory conditions and studied
in standardized situations, were used for this purpose.
Observations showed that normal male and female monkeys
differ markedly in the frequency with which they display
five distinct types of social behavior: (1) threat;

(2) play initiation; (3) rough-and-tumple play; (4)
pursuit play; and (5) mounting behavior. When genetic
females, treated with testosterone propionate from the
39th through the 70th or 105th day of gestational age
and consequently hermaphroditic, are studied under
comparable conditions, their average frequency of performance is intermediate to males and normal females
on every measure.

152. Goy, R.W., Wallen, K. & Goldfoot, D.A. Social factors
affecting the development of mounting behavior in male rhesus
monkeys.   In W. Montagna and W.A. Sadler (Eds.), Reproductive
behavior. New York: Plenum Press, 1974.

The experiments discussed in this paper show that if
opportunities to interact continuously with peers are
provided as late as one year of age, marked improvement
in sexual mounting behavior can result. The authors
have unpublished data showing that continuous social
experience with peers can still be effective when it is
provided at the end of the second year of life. The
majority of males reared with late socialization
procedures were found to be successful copulators in
adulthood. These results stand in contrast to previous
unsuccessful attempts by others to improve sexual
behavior by socialization of isolate-reared males at
relatively advanced ages.

153. Green, P.C. Influence of early experience and age on
expression of affect in monkeys. Journal of Genetic Psychology,
1965, 106, 157-171.

Four groups of monkeys, differing in early experience
and age were trained to accept food readily in an
approach-avoidance situation. Each S was exposed to
10 separate sets of stimuli representing animal, eye,
and human replicas, with the stimuli within each set
presented progressively along a continuum of increasing
realism. Following presentation of the first sequence
of the 10 sets of stimuli, a short rest period
intervened after which a replication of the stimulus
sequence was made. Analyses were made of the various
classes of affective response elicited from the four
treatment groups by stimuli on each level of realism.
Feral juvenile and adult groups showed substantially

higher levels of aggression toward stimuli than did the
dark-reared and maternally separated groups. The latter
two groups showed pronounced attenuation of behavior in
most of the measures used; however all juveniles showed
a common tendency to demonstrate immature forms of behavior
(crouching, screeching, rocking), while the adult animals
tended either to aggress against stimuli or to withdraw
from them. Differences among groups were attributed to
early experience and maturation.

154. Green, P.C., & Gordon, M. Maternal deprivation: Its
influence on visual exploration in infant monkeys. Science, 1964,
145, 292-294.

Visual exploration was studied in maternally reared and
maternally deprived monkeys. When an animal pressed
a bar an opaque screen was raised providing a brief
view of either of a pair of stimuli. Subjects reared
by their mothers pressed more to see animate than
inanimate objects. With increasing age, the number
of bar-pressing responses decreased for an adult female
stimulus, increased for an age peer and for food, and
remained low for geometric forms and an empty chamber.
Maternally deprived subjects established uniformly low
response levels to all stimuli.

155. Griffin, G.A. The effects of multiple mothering on the
infant-mother and infant-infant affectional systems. Unpublished
doctoral dissertation, Madison Wisconsin, 1966.

The purpose of this research was to investigate the effects
of inconsistent mothering produced by a multiplicity of
mother figures on infant-mother and infant-infant
interactions in rhesus monkeys (Macaca mulatta). The
results found in this study were compared with results
found in studies of a similar nature using human infants
and results of studies employing rats as subjects.
There is support in all of these studies for the conclusion that inconsistency of maternal figures is an
important variable in all these species, which affects
the infant's social behaviors with the mother or the
infant's behavior in a nonsocial novel environment.

156.   Griffin, G.A., & Harlow, H.F.   Effects of three months of total
social deprivation on social adjustment and learning in the rhesus
monkey.   Child Development, 1966, 37, 534-547.

    Rhesus monkeys reared in total social isolation for the
    first 3 months of life were compared with 3-month partial
    social isolates on social and learning behaviors.
    Home-cage observations also were taken over the isolation
    and social testing periods. The 3-month total social
    isolates showed extreme withdrawal when they were removed
    from the isolation chambers, and this withdrawal was
    so severe that one S died of starvation because it refused
    to eat food placed in its cage. The totally socially
    isolated Ss exhibited a drop in oral and manual
    exploration of the cage when removed from isolation and
    showed an increase in the category of self-directed
    orality. The isolates also showed difficulty in adapting
    to new situations. However, no differences in social
    or learning behaviors were found between the 3-month
    total social isolates and their partially socially
    isolated controls.

157.   Hamburg, D.A.   Evolution of emotional responses: Evidence
from recent research on nonhuman primates.   In J.H. Masserman
(Ed.), Science and psychoanalysis, Vol. 12, Animal and human.
N.Y.:   Grune and Stratton, Inc., 1968, 39-54.

    This article summarizes the author's view that emotional
    processes serve motivational purposes in meeting crucial
    adaptive tasks: finding food and water, avoiding predators,
    achieving fertile copulation, caring for the young, and
    training the young to cope effectively with the specific
    requirements of a given environment. Data from some
    of the new field observations and experimental studies
    of nonhuman primates--especially on the behavior of
    young monkeys and apes--are presented and discussed.

158.   Hansen, E.W.   The development of maternal and infant behavior
in rhesus monkeys. Behavior, 1966, 27, 107-149.

    This study traced the development of maternal behavior
    in a group of four adult rhesus female monkeys, and
    compared the development of the behavior of their infants
    with the development of the behavior of infant monkeys
    raised on cloth surrogate mothers. The two groups of

subjects were housed, and their behaviors observed, in
separate playpen situations. Social interactions in
pairs and subsequently in groups of four infants were
observed for the first 15 months of the infant's development. The recorded data placed a primary emphasis on
social behaviors and their development, although
additional aspects of infantile behaviors were measured.
Two states in the expression of maternal behavior were
demarcated: a stage of maternal attachment and protection
which was characterized by intensive infant-mother
contact and maternal care-giving behaviors; and a
transitional stage characterized by the fact that the
care-giving behaviors decreased and negative responsiveness
gradually increased. Comparisons of the mother-raised
and surrogate-raised infants permitted an assessment
of the influence of maternal responsiveness on infant
behaviors.

159.   Hansen, E.W., Harlow, H.F., & Dodsworth, R.O.   Reactions of
rhesus monkeys to familiar and unfamiliar peers. Journal of
Comparative and Physiological Psychology, 1966, 61(2), 274-279.

In an attempt to assess the nature of rhesus monkey's
social responding to familiar and unfamiliar peers,
individuals in various combinations of 4 Ss each were
rated over a 15-wk. period in the playpen situation on
a 24-item inventory of social responding. The sex
ratios in the 3 subgroups were not equal. Both the
overall analyses and the analyses which isolated sex
differences showed that several categories of behavior
occurred with significantly different frequency among
familiar as compared to unfamiliar Ss. With minor
exceptions, the interactions involving familiar Ss were
seen to be more positive than those involving unfamiliar
Ss.

160.   Harper, L.V.   Ontogenic and phylogenic functions of the parent-
offspring relationship in mammals. In D.S. Lehrman, R.A. Hinde, &
E. Shaw (Eds.), Advances in the study of behavior (Vol. 3).
New York: Academic Press, 1970, 75-117.

The paper discusses traditional views of the function
of the parent-offspring relationship in mammalian ontogeny
and phylogeny. Various animal species are viewed using
an ethological and anthropological perspective. The

aspects of primate parent-offspring relationships are
discussed from this orientation.

161. Harlow, H.F. The nature of love. The American Psychologist,
1958, 13(12), 673-785.

The maternal deprivation and surrogate mother research
is presented in a general philosophical manner. The
importance of contact is stressed in this article with
implications for several species of animals. The major
portion of the paper deals with infant inadequacy following
maternal deprivation in rhesus monkeys (Macaca mulatta)
as well as the drive for "contact comfort" over other
more vegetative motives.

162. Harlow, H.F. Basic social capacity of primates. Human Biology,
1959, 31, 40-53.

This article presents an extensive review of the literature
dealing with the development of social skills between
mother and infant primates. Early papers by Watson (1914),
Kohler (1925), Yerkes (1929), Harlow (1944), and Butler
(1953, 1954) are discussed with regard to infant
development in both old world monkeys and chimpanzees
under various conditions of maternal deprivation, pseudo-
mothering, and social isolation. The results presented
are consistent with early hypotheses equating maternal
contact with normal psychological and physiological
development in primates as well as other mammalian forms.

163. Harlow, H.F. The development of learning in the rhesus
monkey. American Scientist, 1959, 47, 459-479.

Rhesus monkeys (Macaca mulatta) were tested for learning
capabilities from birth to intellectual maturity.
Various tasks involving spatial discrimination, object
discrimination, delayed response, Hamilton Perseverance
Test, and oddity test indicate that the monkey is
capable of solving simple learning problems during the
first few days of life and that this capability improves
with maturity. Thus, the primate is able to master more
complex problems as a factor of age for the first five
years.

164. Harlow, H.F. Love in infant monkeys. <u>Scientific American</u>,
1959, <u>200</u>, 68-74.

> The affectional systems of infant rhesus monkeys (<u>Macaca</u>
> <u>mulatta</u>) were observed under conditions of maternal
> deprivation and surrogate mother substitution. Surrogate
> mothers constructed of either cloth or wire were placed
> in the infant's cage. The cloth mother provided only
> a soft contact surface while the wire mother allowed
> the infant to derive nourishment via a nipple. The
> classic results demonstrated the importance of contact
> to the developing infant, a need which apparently
> supercedes the need for nourishment based attachment.

165. Harlow, H.F. Affectional behavior in the infant monkey. In
M.A.B. Brazier   (Ed.), <u>The central nervous system and behavior</u>.
Josiah Macy Jr., Foundation Publication, 1960, 307-357.

> The ontogenic development of affectional behavior in
> infant rhesus monkeys (<u>Macaca mulatta</u>) is discussed
> with regard to social isolation and the use of
> surrogate mothers. Special emphasis is placed upon
> central nervous system pathology and behavior displayed
> by the infants. The article presents a general examination
> of primate isolation research and analyzes tangential
> investigations by other scientists.

166. Harlow, H.F. Of love in infants. <u>Natural History</u>, 1960,
<u>69</u>, 18-23.

> Infant rhesus monkeys (<u>Macaca mulatta</u>) were separated from
> their mothers at birth and placed with a wire lactating
> "surrogate mother" or cloth nonlactating "surrogate
> mother". The results of this general article are
> discussed with regard to the development of affectional
> responses in infants as well as the need for contact
> and security.

167. Harlow, H.F. Primary affectional patterns in primates.
<u>American Journal of Orthopsychiatry</u>, 1960, <u>30</u>, 676-684.

> Infant rhesus macaques (<u>Macaca mulatta</u>) serve as a model
> for the development of affectional systems in monkeys,

apes, and man. Five affectional systems are delineated,
these are: infant for mother, infant for peer, heterosexual, maternal, and paternal. The research conducted
with surrogate mothers provides the foundation for this
article.

168. Harlow, H.F. The development of affectional patterns in
infant monkeys. In B.H. Foss (Ed.), Determinants of infant
behavior. London: Methuen and Co., Ltd., 1961, 75-97.

   The article discusses the important reflexes of infant
   rhesus macaques (Macaca mulatta) which play an important
   role in the affectional ties between infant and mother.
   Infants were separated from their mothers at birth and
   raised with either a cloth surrogate (lactating or
   nonlactating) or a wire surrogate (lactating or nonlacting)
   under moving or stationary conditions. The results
   of the investigation are directed toward the development
   of affectional responses between age-mates and heterosexual partners.

169. Harlow, H.F. Development of affection in primates. In E.
Bliss (Ed.), Roots of behavior: Genetics, instinct, and socialization in animal behavior. New York: Harper (Hoeber), 1962,
157-166.

   The article presents an analysis of the development of
   affectional stages in infant rhesus monkeys (Macaca mulatta)
   under normal and surrogate mother conditions (cloth
   nursing, cloth nonnursing, wire nursing, wire nonnursing,
   stationary, or moving). The results of the experimentation
   indicate that initial contact is reflexive in nature,
   but more complex affectional and manipulatory behavior
   is dependent upon reciprocal interaction with a peer
   or parent.

170. Harlow, H.F. Development of second and third affectional
systems in macaque monkeys. In T.T. Tourlentes, S.L. Pollack,
and H.E. Himwich (Eds.), Research approaches to psychiatric
problems. New York: Grune and Strattaon, 196?, 209-229.

   The article discusses the research conducted with infant
   rhesus macaques (Macaca mulatta) involving the development

of affectional responses following maternal deprivation
and surrogate mother conditions. Infants were observed
in a social playroom with other monkeys and objects of
exploration following removal from the surrogate mother.
The results of this research are discussed with regard
to infant-infant affectional development and heterosexual
behaviors.

171. Harlow, H.F. The effects of radiation on the central nervous
system on behavior-general survey. Responses of the nervous system
to ionizing radiation. New York: Academic Press, 1962, 627-644.

This article presents a general discussion of the effects
of radiation upon the CNS of a variety of animals including
macaque monkeys. Various measures of learning (oddity
and visual curiosity tests), social adjustment, and
taste aversion are discussed with regard to levels of
radiation and physiological variables.

172. Harlow, H.F. The heterosexual affectional system in monkeys.
American Psychologist, 1962, 17, 1-9.

This article presents a general analysis of the heterosexual affectional system in the rhesus monkey (Macaca
mulatta). A series of developmental stages (infantile
heterosexual, preadolescent, adolescent, and mature)
are discussed with regard to normal and surrogate mother
conditions. All animals were observed in a large playpen enclosure with a conspecific of the opposite sex.
The results of the investigation are discussed with
regard to sociopathic syndromes and possible therapeutic
techniques for heterosexual inadequacy seen in animals
deprived of maternal contact.

173. Harlow, H.F. Mothers, monkeys, and sex. Sexology, 1962,
4-10.

The article presents an array of information dealing
with the sexual behavior of infant rhesus macaques
(Macaca mulatta) exposed to surrogate mothers rather
than their natural mothers. Development of sexual
behaviors are traced under both natural and surrogate
conditions. Infants raised by cloth surrogate mothers

appear to be devoid of natural social behaviors and
respond inappropriately to advances made by other
conspecifics.

174. Harlow, H.F. Basic social capacity of primates. In C.H.
Southwick (Ed.), Primate social behavior. New York: VanNostrand
Reinhold Co., 1963, 153-160.

Harlow presents a broad discussion of the social and
intellectual development of primates incorporating
naturalistic observations with experimental findings.

175. Harlow, H.F. The maternal affectional system. In B.M. Foss
(Ed.), Determinants of infant behavior (Vol. 2). London: Methuen
and Co., 1963. Proceedings of a Tavistock study group held in
the house of the CIBA Foundation, London, 1961. New York:
John Wiley and sons, 3-33.

This paper presents a description and analysis of the
maternal affectional system through which the mother
rhesus monkey (Macaca mulatta) relates to her own offspring and other infants. Information regarding the
development of this affectional system is derived from
a playpen situation under natural and surrogate mother
conditions. Aspects of the Bowlby syndrome are discussed
with emphasis on maternal separation and reunion.

176. Harlow, H.F. Early social deprivation and later behavior in
the monkey. In A. Abrams, H.H. Gardner, and J.E.P. Toman (Eds.),
Unfinished tasks in the behavioral sciences. Baltimore: Williams &
Wilkins, 1964, 154-173.

Harlow discusses the effects of various kinds of early
social deprivation upon later behavior in monkeys.
The effects of inadequate mothering, absence of mothering,
and limited or controlled social experience are investigated.
The research involves separating baby monkeys from their
mothers at birth and raising them under total social
deprivation, partial social deprivation, or under
controlled social stimulation. Isolates were observed
both during and following isolation. Post isolation
observations were carried out in a playpen enclosure
with a wide array of stimuli, including: lactating and
nonlactating surrogate mothers (wire and cloth respectively),
toys, and other infants.

177. Harlow, H.F. Sexual behavior in the rhesus monkey. In F.A.
Beach (Ed.), <u>Sex and behavior</u>. New York: Wiley, 1965, 235-265.

The central topic of this article is the development of
sexual behavior in the rhesus monkey (<u>Macaca mulatta</u>).
These behavior patterns are discussed developmentally
in feral animals and animals raised under a variety of
laboratory conditions, i.e., separating infants from
their mothers at birth and subsequently raising them
in hardware cloth cages or with various kinds of
surrogate mothers.

178. Harlow, H.F. Total social isolation: Effects on macaque
monkey behavior. <u>Science</u>, 1965, <u>148</u>, 666.

Sixteen rhesus monkeys were separated from their mothers
at birth. Twelve were placed in isolation chambers for
3, 6, or 12 months, during which period they saw no
human being or animal. Four monkeys were raised in semi-
isolation for 6 months and then in isolation for 6 months;
in semi-isolation they were given extensive social
experience in a playroom, in pairs with a pair of equal-
aged, semi-isolated, control monkeys, to examine their
social behavior. The 3-month isolates, after recovery
from initial shock, made effective social contacts with
controls and with each other. The 6-month isolates failed
to adjust to the controls and were enormously impaired
in play with each other. The 12-month isolates failed
completely with controls and with each other. The
impairment of the 6- and 12-month groups appears to be
permanent. The monkeys isolated for 6 months (after
semi-isolation from birth to 6 months) reacted effectively
with controls and with each other in a relatively short
time, but showed excessive aggression. Isolates and
controls were also tested for learning ability on an
extensive battery of problems. The differences slightly
favored the control groups although in no instance were
the differences statistically significant. Results
indicate that monkeys can withstand at least 3 months
of social isolation starting at birth, or 6 months of
total social isolation starting at 6 months of age,
but their social potentials are destroyed if isolation
from infancy persists for 6 to 12 months. Learning
ability is apparently not impaired by isolation.

179. Harlow, H.F. The primate socialization motives. <u>Transactions</u>
<u>and Studies of the College of Physicians of Philadelphia</u>, 1966,
<u>33</u>, 224-237.

This article discusses the effects of partial and total
social isolation on the development of affectional responses
and social behavior in infant rhesus macaques (<u>Macaca</u>
<u>mulatta</u>). The five affectional systems discussed are:
the mother-infant system, the infant-mother system, age-
mate system, heterosexual system, and the paternal
affectional system. Observations of the isolated infants
in a playroom condition led to the conclusion that lack
of contact leads to decrements in social behavior and
an absence of affectional responses.

180. Harlow, H.F. Age-mate or peer affectional system. In
D.S. Lehrman, R.A. Hinde, and I. Shaw (Eds.), <u>Advances in the study</u>
<u>of behavior</u> (Vol. 2). New York: Academic Press, 1969, 333-383.

The article presents an in depth examination of the
age-mate or peer affectional system in primates. The
development of stages of intimate object play (reflex,
exploratory, social play, and aggressive play) are
examined under various social conditions, i.e.,
partial social isolation, total social isolation, mother
and peer raised, and motherless-mother raised. Infant
rhesus monkeys (<u>Macaca mulatta</u>) were then observed in
an open playroom condition. The results presented
depict the isolate infant as socially inadequate,
depressed, and hyper-or hypo-aggressive.

181. Harlow, H.F. Love created-love destroyed-love regained. In
<u>Modeles animaux du compart humain</u>. Paris: Editions du Centre
National de la Recherche Scientifique. 1972, <u>198</u>, 13-60.

This general review article discusses the classic research
of affectional development in the infant rhesus monkey
(<u>Macaca mulatta</u>) exposed to various levels of social
stimulation and deprivation.

182. Harlow, H.F. Induction and alleviation of depressive states
in monkeys. In N.F. White (Ed.), <u>Ethology and psychiatry</u>. (From
the Clarence M. Hicks Memorial Lectures, McMaster University, 1970)
Toronto: University of Toronto Press, 1974, 197-208.

During the last two years we have initiated a research
programme designed to induce or simulate various forms
of human depression in rhesus monkeys (<u>Macaca mulatta</u>).
If monkeys can be made depressed for a prolonged period
of time a variety of biological and social studies
become possible. Prior to this particular programme,
we had already conducted two studies of infantile
anaclitic depression produced by separating infant
monkeys from their mothers at six months of age (Seay,
Hansen, and Harlow, 1962). This research had served
as a model for other investigators testing the effects
of infant separation from the mother and mother
separation from the infant or rhesus, bonnet, and
pigtail macaques.

183. Harlow, H.F. Maternal and peer affectional deprivation in
primates. In J.H. Cullen (Ed.), <u>Experimental behavior: A basis
for the study of mental disturbance</u>. Dublin: Irish University
Press, 1974, 85-98.

Experimental maternal deprivation in primate infants is
a devastating event because it destroys the intimate
love bonds between mother and infant. This lack of
maternal contact leads to a wide variety of developmental
disorders which are analogous to those seen in autistic
or severely depressed children. A surrogate mother made
of cloth may satisfy basic needs for contact, however
social and sexual inadequacy is still quite prevalent
among partially deprived infants. If a human or monkey
infant is seperated from its mother for a period of
days anaclitic depression will result. Three stages
of this depression are formulated by Bowlby, they are:
protest, dispair, and detachment. Early depression
often involves loss of social bonds, however, juvenile
depression may be produced by multiple separations,
or more dramatically by confinement in a vertical chamber.
Once depression has been established as a syndrome therapy
is initiated in an attempt to ameliorate the bizarre
behavior patterns. Successful behavior rehabilitation
has been achieved in monkeys so abnormal that they were
thought to be doomed to social extinction. Thus,
the full gambit of illness and treatment has been
modeled using infant rhesus monkeys raised under a
variety of deprivation conditions.

184. Harlow, H.F. and Griffin, G. Induced mental and social deficits
inrhesus monkeys. In S.F. Osler & R. Cook (Eds.), The biosocial basis
of mental retardation. Baltimore: The Johns Hopkins Press, 1965,
87-106.

   Infant rhesus monkeys (Macaca mulatta) raised under various
   degrees of maternal and peer deprivation were observed
   for detriments in social and intellectual development.
   The results of these investigations are paramount to an
   understanding of maternal inadequacy and social ineptness
   seen in adult isolates. In addition, a brief introduction
   on induced biochemical insufficiencies in infant monkeys
   provides a unique preface for this article.

185. Harlow, H.F. and Harlow, M.K. A study of animal affection.
Natural History, 1961, 70, 48-55.

   This general article presents a wide array of infant rhesus
   monkey (Macaca mulatta) developmental research conducted
   at the University of Wisconsin Primate Laboratory.
   The infant-infant and infant-maternal affectional
   systems are discussed with regard to normal and isolation
   conditions. The results of these investigations demonstrate
   the importance of contact and companionship with either
   peers, natural mother, or surrogate mother.

186. Harlow, H.F. and Harlow, M.K. The effect of rearing conditions
on behavior. Bulletin of the Menninger Clinic, 1962, 26, 213-224.

   Infant rhesus monkeys were reared under a variety of
   social conditions, including total isolation and partial
   isolation. The macaques were maintained in either
   individual bare wire cages housed in a colony room for
   at least the first two years of infancy, or in individual
   wire cages with access to one or two mother surrogates
   for the first six months of life. The other conditions
   included situations with real or surrogate mothers
   plus contact with other infants for the first year or
   two of life. Total isolation for two years resulted
   in social impairment leading to lack of appropriate
   social displays and sexual maladjustment for the next
   two years of communal living. Partial isolation yielded
   behavioral abberations in many monkeys and sexual inadequacy
   in all males and all but one female. Four impregnated
   females proved to be completely inadequate mothers.

Infants raised by surrogate mothers were less advanced
in social skills than their natural mother counterparts.
Over all, the more complete and lengthy the social
deprivation, the more devastating the behavioral effects.

187. Harlow, H.F. and Harlow, M.K. Social deprivation in monkeys.
Scientific American, 1962, 207, 136-146.

Infant rhesus macaques (Macaca mulatta) were exposed to
various forms of social isolation and maternal separation
and observed for social impairment. The article presents
a general review of this research with an extensive analysis
of developmental psychopathology and supportive therapy
for deprived infants using peer contact.

188. Harlow, H.G. and Harlow, M.K. The affectional systems. In
A.M. Schrier, H.F. Harlow, and F. Stolintz (Eds.), Behavior of
nonhuman primates (Vol. 2). New York: Academic Press, 1965,
287-334.

We are convinced that there are at least five affectional
systems in the primate order: the systems of infant-
mother affection, mother-infant affection, peer affection,
heterosexual affection in adults, and paternal affection.
We believe these systems all go through an orderly series
of maturational stages and we also believe that they
operate through different behavioral, neural, and biochemical variables. Some of the systems, such as the
mother-infant and the infant-mother system, depend on
similar variables, but in other systems, such as the
age-mate or peer affectional system, the variables differ
strikingly from those found elsewhere.

189. Harlow, M.K. & Harlow, H.F. An analysis of love. The Listener,
1965, 73, 255-257.

The article discusses the five primate affectional systems:
infant-mother, mother-infant, age-mate, heterosexual,
and paternal.

190. Harlow, H.F. and Harlow, M.K. The effects of early social
deprivation on primates. In J. de Ajuriaguerra (Ed.),
Desafferentation experimentale et clinique. Geneva, Switzerland:
George E. Cie S. A., 1965, 67-77.

    The social development of rhesus monkeys has been studied
    extensively at the University of Wisconsin Primate
    Laboratory, and a number of experiments have imposed
    deprivations of social experience on infants. Total
    social isolation from birth to three months of age
    produces reversible effects, but six months or twelve
    months of this treatment creates permanent social deficits
    while leaving the intellectual functions unimpaired.
    Partial social isolation in which monkeys are deprived
    from birth to six months or twelve months of age of all
    physical contacts with their kind but still see and
    hear other monkeys leaves personal, sexual, and social
    deviations in both sexes. Some females eventually make
    an adequate sexual adjustment with experience, but no
    male has adjusted sexually in spite of years of pairings
    with females in estrus. No mothering, artificial mothering,
    or cruel or indifferent mothering affects early adjustments
    to peers in infants permitted regular play experience
    with age mates from early life, but social and sexual
    behavior by the second year is adequate for the first
    group and normal in the latter two groups. Mothering
    without peer experience until eight months of age
    produces monkeys with adequate sexual adjustment but
    hyperaggressive and fearful of physical contact. Four
    months of confinement with mothers produces less drastic
    deviations in these respects, but the animals are
    somewhat more aggressive and fearful of controls having
    mothering from birth and peer experience after the first
    few weeks of life.

191. Harlow, M.K. and Harlow H.F. Romulus and rhesus. The
Listener, 1965, 73, 215-217.

    This article discusses the effects of social isolation
    on infant rhesus monkeys (Macaca mulatta). The
    developmental repercussions of this research are
    presented in a general way with an emphasis on human
    implications.

192. Harlow, H.F. and Harlow, M.K. Learning to love. American
Scientist, 1966, 54(3), 244-272.

This paper discusses three primate affectional systems:
the mother-infant affectional system, the infant-mother
affectional system, and the age-mate or peer affectional
system. These stages of development are traced in the
rhesus monkey (Macaca mulatta) under various levels
of social deprivation and surrogate mothering.

193. Harlow, H.F. and Harlow, M.K. The young monkeys. Psychology
Today, 1967, 1, 41-47.

The article presents a very general overview of primate
affectional systems and social communication under normal
and isolation conditions.

194. Harlow, H.F., and Harlow, M.K. Effects of various mother-
infant relationships on rhesus monkey behaviors. In B.M. Foss (Ed.),
Determinants of infant behavior (Vol. 4). London: Methuen,
1969, 15-36.

This article presents a general summary of the research
dealing with mother-infant rearing relationships and the
development of social behaviors in rhesus monkeys (Macaca
mulatta). Monkeys were raised from birth without mothers
or playmates or raised with mothers and easily available
playmates. Intermediate rearing conditions included
normal mothering without playmates and a variety of
situations in which playmates were available, but mothering
ranged from total maternal deprivation, to inanimate,
nonresponsive surrogate mothers, indifferent or brutal
mothers, and a rotating series of individual mothers.

195. Harlow, H.F., and Harlow, M.K. Developmental aspects of
emotional behavior.   In P. Black (Ed.), Physiological correlates
of emotion. New York: Academic Press, 1970, 37-58.

As the data have accumulated on many groups of infant
monkeys subjected to varying rearing conditions, we have
come to be impressed by alternative routes monkeys may
take to achieve adequate social behavior, which by our
criteria would include attachment to peers, control of

fear and aggression, and normal sexual patterns. Under
the protected conditions of the laboratory, peer socialization and mother-infant socialization appear in large
part to be interchangeable in the development of the
infant. Learning is commonly conceived as a process that
shapes preestablished, unlearned response patterns.
This is only part of the picture insofar as social
learning is concerned. One of the most powerful functions
of early social learning in primates, and possibly in
all mammals and in many other classes of animals as well,
is that of developing social patterns that will restrain
and check later maturing behaviors having an asocial
potential. It is the establishment of positive, learned
social patterns before negative, unlearned patterns
emerge. It is in this sense an anticipatory form of
learning, a check against the inappropriate exercise
of negative behavior patterns within the social group
while permitting their appropriate expression toward
threatening intruders from without.

196. Harlow, H.F. and Harlow, M.K. Psychopathology in monkeys.
In H.D. Kimmel (Ed.), Experimental psychopathology. New York:
Academic Press, 1971, 203-229.

The development of psychopathology in infant rhesus monkeys
(Macaca mulatta) seen following maternal deprivation is
discussed with particular emphasis on anaclitic depression.
The major emphasis of the manuscript is the importance
of contact for the normal development of the infant monkey.
Various types of maternal and social isolation are
discussed, including: surrogate mothers (cloth or wire,
lactating or nonlactating), partial social isolation,
total social isolation, peer deprivation, and pit-
produced depression. A hierarchy of pathological
symptomology are analyzed relevant to the degree of
isolation, age of the isolate, and period of isolation.
A final summary provides a brief overview of future
trends in this area of research.

197. Harlow, H.F. and Harlow, M.K. The language of love. In
T. Alloway, L. Krames, & P. Pliner (Eds.), Communication and
affect. New York: Academic Press, 1972, 1-18.

This article presents a thorough analysis of the development of affectional systems within rhesus monkeys (Macaca

mulatta) under normal and surrogate mother conditions.
The important aspects of social communication are discussed
with regard to proper affectional development.

198. Harlow, H.F., and McKinney, W.T., Jr. Nonhuman primates and
psychoses. Journal of Autism and Childhood Schizophrenia, 1971,
1, (4) 368-375.

      Studies using nonhuman primates have facilitated our
      understanding of human psychopathology and in particular
      have provided some models of abnormal behavior occurring
      in the young, developing organism. The theoretical
      linkages between abnormal behavior in rhesus monkeys
      and in human beings are discussed. Two research areas
      are cited as examples where experiments with monkeys
      have provided some reasonable models for human
      psychopathology. These two areas are total social
      isolation and disruption of affectional bonds between
      mothers and infants or between peers. Finally, the
      philosophical issues concerning the production of
      experimental psychopathology in animals are discussed
      and criteria presented to guide future research in
      this area.

199. Harlow, H.F., & Novak, M.A. Psychopathological perspectives.
Perspectives in Biology and Medicine, 1973, 16(3), 461-478.

      This article presents an analysis of psychopathology
      in rhesus monkeys (Macaca mulatta) induced through
      various forms of social isolation. The isolation
      syndromes are discussed interms of behavioral abnormal-
      ities including: stereotypies, self-clasping, rocking,
      huddling, and lack of adequate social skills in the
      presence of normal animals.

200. Harlow, H.F. & Rosenblum, L. Maturational variables
influencing sexual posturing in rhesus monkeys. Archives of
Sexual Behavior, 1971, 1, 73-78.

      Measurement of the social behaviors of rhesus monkeys
      (Macaca mulatta) yielded three distinct primary patterns
      of sex-differentiating behaviors, these are: threat,
      passivity, and rigidity. These three patterns are

clearly sex-differentiating and become progressively more
sex-differentiating with maturity. The communicative
value of these behaviors is clearly representative
of dominance status or appeasement responses to dominance
expressed by other primates. The threat response is
indicative of a positive dominant behavior, and both
passivity and postural rigidity are expressions of
submissive behavior or sexual compliance.

201. Harlow, H.W. & Seay B. Affectional systems in rhesus monkeys.
Journal of the Arkansas Medical Society, 1964, 61(4), 107-110.

An array of research investigations conducted at the
University of Wisconsin Primate Laboratory were designed
to assess the effects of various kinds of social rearing
conditions on subsequent social, sexual, and maternal
adjustment of rhesus monkeys (Macaca mulatta). Groups
of monkeys were raised in total social isolation from
birth until 3 months, 6 months, and 12 months of age.
The results of these investigations indicate that there
is some crucial period between the sixth and twelfth
month which is critical for normal social development.

202. Harlow, H.F., & Seay, B. Mothering in motherless mother
monkeys. British Journal of Social Psychiatry, 1966, 1, 63-69.

The maternal behaviour of four severely socially restricted
rhesus monkeys was described and compared to that of four
captive feral rhesus monkeys. All four macaque mothers
suffering severe social restriction during the first 18
months of life were totally inadequate and none of their
infants would have survived had not the experimenters
intervened. Two of this initial motherless mother
group were primarily abusive and two primarily indifferent
to their babies. Second infants of these motherless
mothers have so far been accorded normal maternal
care. Supplementary data from five additional motherless
mothers and their infants were also reported. The early
life history of these mothers was not uniform and playpen
groups could not be formed because of wide infant age
discrepancies.

203. Harlow, H.F. & Suomi, S.J. Induced psychopathology in monkeys.
Engineering and Science, California Institute of Technology, 1970,
33, 8-13.

This paper presents the research performed at the University
of Wisconsin Primate Laboratory on developmental psychopathology in rhesus monkeys (Macaca mulatta). The effects
of partial and total social isolation, maternal deprivation or separation, and immersion in the vertical chamber
are discussed with regard to the development of depressive
states in nonhuman primates. The implications of this
research are directed toward human psychopathology and
modes of treatment.

204. Harlow, H.F. & Suomi, S.R. Induction and treatment of
psychiatric states in monkeys. Proceedings of the National
Academy of Science, 1970, 66, 241.

Four rhesus monkeys were subjected to total social
isolation from birth until 6 months of age. Previous
research has shown that this period of privation produced
monkeys that could not adjust socially to age-mates; were
grossly sexually deficient, particularly in males;
became indifferent or brutal mothers if impregnated;
and remained socially fearful and nonresponsive as adults.
In spite of pervasive social fear the deprived adults
were abnormally aggressive toward helpless infants.
Intensive tests showed that these adult rhesus suffered
little or no learning or intellectual loss. At 6 months
the deprived monkeys in this study were removed from
isolation and placed in cages adjacent to four 3-month
old "psychiatrist" monkeys who had been surrogate-
reared with 2 hours of daily peer social interaction
and who exhibited essentially normal exploratory and
social behavior both in social and nonsocial situations.
The isolate monkeys were then allowed to interact with
the two psychiatrist monkeys 2 hours per day in two
situations: in pairs within the home cage and as a group
of four in a larger playroom. Within 3 weeks all four
isolates showed dramatic improvement in both social and
nonsocial home cage behavior as indicated by decreases
in self-orality, self-clasping and rocking, huddling,
and by increases in exploratory and locomotor behavior
as well as by rapid emergence of social contact and play.
Similar dramatic recovery, although slightly more
delayed, was observed in three of the four isolate
subjects in the playroom situation.

205. Harlow, H.F. & Suomi, S.J. Nature of love-simplified.
American Psychologist, 1970, 25-(2), 161-168.

   The surrogate mother is thoroughly investigated as a
   substitute mother for the infant rhesus monkey (Macaca
   mulatta). The basic orientation of the research is to
   analyze various surrogate variables, i.e., lactation,
   facial characteristics, body-surface characteristics,
   motion components, and temperature variables with regard
   to infant preference and development.

206. Harlow, H.F. & Suomi, S.J. Production of depressive behaviors
in young monkeys. Journal of Autism and Childhood Schizophrenia,
1971, 1(3), 246-255.

   Socially unsophisticate rhesus monkeys ranging in age from
   6 to 13 months were individually confined in vertical
   chambers for a total of 30 days. Upon emergence from
   the chambers they exhibited significant increases in
   self-clasping and huddling behaviors and decreases in
   locomotive and exploratory activity. These behavioral
   changes were similar in direction to those resulting
   from separation from attachment objects, which have
   been described as depressive. The utility of the vertical
   chamber apparatus as a device for the production and
   study of depression in monkeys is discussed, and implications for the study of child psychoses noted.

207. Harlow, H.F. & Suomi, S.J. Social recovery by isolation-reared
monkeys. Proceedings of the National Academy of Science, 1971,
68(7), 1524-1538.

   Total social isolation of macaque monkeys for at least
   the first 6 months of life consistently produces severe
   deficits in virtually every aspect of social behavior.
   Experiments designed to rehabilitate monkeys reared in
   isolation are described. While young isolates exposed
   to equal-age normal peers achieved only limited recovery
   of simple social responses, some mothers reared in
   isolation eventually exhibited acceptable maternal
   behavior when forced to accept infant contact over a
   period of months, but showed no further recovery; isolate
   infants exposed to surrogates were able to develop crude
   interactive patterns among themselves. In contrast to
   the above results, 6-month-old social isolates exposed

to 3-month-old normal monkeys achieved essentially
complete social recovery for all situations tested.
It is postulated that social stimulation that both
permits subjects to achieve contact acceptability and
provides an interactive medium conducive to gradual
development of sophisticated social behaviors will result
in almost complete recovery of social capabilities
previously obliterated by rearing in isolation.

208.   Harlow, H.F. & Suomi, S.J.   Induced   depression in monkeys.
Behavioral Biology, 1974, 12, 273-296.

Recent work at the Wisconsin Primate Laboratory has been
directed toward experimental production and cure of
human-type psychopathologies in rhesus monkey subjects.
A primary emphasis has been on depression.   Data are
presented which identify different procedures designed
to produce depression, help determine the susceptibility
of any given subject to such procedures, and which outline
methods and modes of therapy.   The significance and
future of work in the area of monkey models of human
disorders is discussed.

209.   Harlow, H.F. & Suomi, S.J.   Workshop I: Instinctual and
environmental learning.   In G. Serban and A. Kling (Eds.),
Animal models in psychobiology.   New York: Plenum Press, 1976,
61-74.

This workshop examines the use of animal models, particularly those based on studies with nonhuman primates,
to facilitate the understanding of several areas of human
psychobiology.

210.   Harlow, H.F. & Zimmerman, R.R.   The development of affectional
responses in infant monkeys.   Proceedings of the American
Philosophical Society, 1958, 102(5), 501-509.

Infant rhesus monkeys were separated from their mothers
six to twelve hours after birth and were housed in
individual cages.   A cloth surrogate mother and a wire
surrogate mother were placed in different cubicles
attached to the infants living environment.   Four infant
monkeys experienced a cloth mother which lactated and

a wire mother which did not. The second set of infants
experienced the exact opposite conditions with a lactating
wire mother and a cloth mother which did not. In either
condition the infant received all its milk through the
mother surrogate as soon as it was able to maintain
itself in this manner. Thus, the experiment was designed
to test the relative importance of the variables of
contact comfort and nursing comfort. The results of
the study demonstrated the overwhelming importance of
contact comfort in the development of affectional responses,
whereas lactation is a variable of negligible importance
when compared to the aforementioned variable, i.e., contact.

211. Harlow, H.F. & Zimmermann, R.R. Afectional responses in
the infant monkey. Science, 1959, 130, 421-431.

   Sixty infant rhesus macaque monkeys (Macaca mulatta)
   were separated from their mothers 6 to 12 hours after
   birth and raised with cloth or wire surrogate mothers.
   These surrogate mothers were of the lacatating or nonlactating variety and allowed infants to suckle and
   cling to the ventral surface while nursing. The results
   of this investigation are discussed with regard to the
   development of affectional and social behaviors.

212. Harlow, H.F., Akert, K. and Schiltz, K.A. The effects of
bilateral prefrontal lesions on the learned behavior of neonatal,
infant, and preadolescent monkeys. The frontal granular cortex
and behavior. New York: McGraw-Hill, 1964, 126- 148.

   Extensive bilateral prefrontal lesions were produced in
   rhesus monkeys at 5 days of age, before delayed-response
   learning is demonstrable; at 150 days, when the ability
   has matured to some extent; at one year, when the ability
   is advanced but short of adult status; and at two years,
   when the function is at or close to maximal level. All
   animals were subjected to extensive behavioral testing
   postoperatively and compared in performance with normal
   animals of approximately the same ages. The test battery
   included discrimination learning, delayed response,
   string tests, discrimination learning set, oddity learning
   set, and the Hamilton search test. It was predicted
   that deficits would be confined to delayed-response
   performance, the 5-day group showing no loss, the 150-
   day group mild loss, and the two older groups showing

serious and complete loss, respectively. It was also
expected that the hyperactivity and circling behavior
pattern appearing in adults after prefrontal ablation
would also occur in the two older operated groups.
Results were in accord with predictions except that there
was no delayed-response deficit in the 150-day group
and there was no hyperactivity pattern in the one-year
group. A loss for the two-year group on the Hamilton
search test is uncertain but is a possibility that needs
further investigation for verification.

213.   Harlow, H.F., Blazek, N.C., & McClearn, G.E.   Manipulatory
motivation in the infant rhesus monkey.   Journal of Comparative
and Physiological Psychology, 1957, 49, 444-448.

The manipulatory behavior of six infant rhesus monkeys
was investigated during a series of four problems measuring
motivation and learning based on manipulation motives.
Testing was initiated when the Ss were between 16 and
36 days of age and before the introduction of solid food.
Manipulation was found generally to increase in amount
and efficiency with age and with practice. The results
suggest that manipulatory behavior is self-sustaining
and is not dependent upon, nor derived from, internal
drives such as hunger or thirst, or their incentive
systems.

214.   Harlow, H.F., Dodsworth, R.O., & Harlow, M.K.   Total social
isolation in monkeys.   Proceedings of the National Academy of
Science, 1965, 54, 90-97.

The findings of the various total and partial isolation
studies of rhesus monkeys (Macaca mulatta) suggest that
severe and enduring social isolation leads to social
stagnation characterized by emotional instability,
infantile behaviors, and sexual inadequacy. In contrast,
short periods of total social isolation leave no permanent
deficits in social adjustment. Thus, it appears that
there are critical periods for social stimulation and
subsequent behavioral stability.

215. Harlow, H.F., Harlow, M.K., & Hansen, E.W. The maternal
affectional system of rhesus monkeys. In H.L. Rheingold (Ed.),
Maternal behavior in mammals. New York: Wiley, 1963, 254-281.

    The maternal affectional system is one of a number of
    affectional systems exhibited by rhesus monkeys (Macaca
    mulatta). It has been studied extensively by observing
    rhesus mothers with their offspring from birth until
    18 or 21 months of age in playpen situations. Maternal
    behavior in the rhesus monkey is characterized by
    sequential stages designated as attachment and protection,
    ambivalence, and separation or rejection. Several factors
    influence the development and expression of maternal-
    infant affection, these include: the mother's developmental
    background, contact or availability of peers, separation
    or other stressful events, and the number of previous
    pregnancies and surving offspring ascribable to the
    mother.

216. Harlow, H.F., Harlow, M.K. & Suomi, S.J. From thought to
therapy: Lessons from a primate laboratory. American Scientist,
1971, 59, 538-549.

    This article presents a thorough analysis of the development
    of behavioral abnormalities in infant rhesus monkeys
    (Macaca mulatta). The research discussed includes:
    surrogate mothers, total social isolation, partial social
    isolation, and maternal separation. Basic affectional
    systems are also discussed with regard to these conditions
    and natural maternal and peer stimulation. The treatment
    of psychopathology and depression are analyzed using
    this primate model.

217. Harlow, H.F., Plubell, P.E., & Baysinger, C.M. Induction
of psychological death in rhesus monkeys. Journal of Autism and
Childhood Schizophrenia, 1973, 3(4), 299-307.

    An experimental study designed to induce psychological
    death in rhesus monkeys is presented and discussed.
    Four infant monkeys were raised with variable temperature
    surrogates. A 20-minute surrogate cold schedule was
    imposed 3 times per day, 5 days per week. All animals
    showed progressively increasing frequencies of disturbance
    behaviors. The introduction of a nightly 12-hour cold
    surrogate period at experimental week 9 rapidly produced

a dramatic increase in disturbance frequency for all
infants, and at the end of 2 weeks appeared to precipi-
tate the death of one subject. A three-fold criterion
of impending psychological death was established and
successfully applied to the infants of a subsequent
study. It is suggested that the case yields further
presumptive evidence that imminent psychological death
produced by social loss can be detected in time to take
appropriate remedial action. Also, this animal model
of extreme depression may determine the important
variables underlying the disorder.

218. Harlow, H.F., Rowland, G.L., & Griffin, G.A. The effect of
total social deprivation on the development of monkey behavior.
Psychiatric Research Reports of the American Psychiatric Association,
1964, 19, 116-135.

The effects of partial social isolation on the social
and sexual behavior of rhesus monkeys have been studied
in detail during the last six years and further analyses
concerning duration of partial deprivation and effects
of early "normal" social experience on subsequent deprivation
are in preparation. The only previous Wisconsin paper
on total social deprivation did not vary the deprivation
interval and antedated our social playroom and playpen
techniques. The present paper details the initiation
of a major research program concerning the effects of
total social deprivation. Groups of rhesus monkeys were
subjected to 3-, 6-, and 12-month periods of total social
isolation from birth onward and to total social isolation
during the second half of the first year of life. The
data indicated increasing socially debilitating effects
with increasing period of total social confinement, but
also suggested that qualitatively different types of social
disorders might be a function of the social deprivation
scheduling. Total social deprivation exerted an adverse
effect in many of the subjects upon unlearned responses,
freezing in the shock avoidance situation, and fear in
the WGTA test apparatus, upon which learned behaviors
were dependent, but intellectual deterioration was not
discovered under the test conditions reported.

219. Harlow, H.F., Schlitz, K.A., & Harlow, M.K. Effects of social
isolation on the learning performance of rhesus monkeys. Proceedings
of the 2nd International Congress of Primatology, 1969, 1, 178-185
(Karger/Basel, New York).

Normal monkeys and monkeys previously subjected to 9
months of social isolation were tested on discrimination
learning, learning set, delayed response, and oddity
learning set. The social isolates were inferior on
the first half of the delayed response problems but not
the second. On all other tests the social isolates
performed at least as effectively as the normal controls.
Limited data on a group of monkeys subjected to 6 months
of social isolation give no indication of any learning
impairment on the tests thus far completed. Similarly,
previous studies do not suggest a learning deficiency
in social isolates. It is the conclusion of the authors
that social isolation produces emotional problems in
rhesus monkeys which make adjustment to social and
learning conditions difficult but leaves their intellectual
functioning unimpaired.

220. Harlow, H.F., Suomi, S.J. & McKinney, W.T., Jr.,   Experimental
production of depression in monkeys. Mainly Monkeys, Newsletter
of the Wisconsin Regional Primate Research Center, 1970, 1, 6-12.

The article discusses the effects of partial and total
social isolation on the development of infant rhesus
monkeys (Macaca mulatta). Emphasis is placed upon the
psychopathology of isolates analogous to symptoms seen in
human depressive patients. The use of the vertical chamber
and other techniques for the development of experimental
psychopathology are discussed.

221. Harlow, H.F., Harlow, M.K., Dodsworth, R.O., & Arling, G.I.
Maternal behavior of rhesus monkeys deprived of mothering and
peer associations in infancy. Proceedings of the American
Philosophical Society, 1966, 110(1), 58-66.

The behavior of 20 "motherless" female rhesus monkeys
(Macaca mulatta) toward their offspring was observed after
the birth of the infants, and the mothers were classified
as abusive, indifferent, or adequate on the basis of
their treatment of the infant during the first month of
life. Observations of the motherless mothers and normal
mothers lead to the conclusion that motherless mothers
are socially inept and seem to show maternal improvement
following second and third pregnancies. The article
provides an extensive analysis of maternal-infant
interactions based upon the mother's developmental
background.

222. Harlow, H.F., Harlow, M.K., Hansen, E.W., & Suomi, S.J.
Infantile sexuality in monkeys. Archives of Sexual Behavior, 1972,
2(1), 1-7.

    Developmental patterns of sexuality for young male and
female rhesus monkeys are illustrated. These patterns
show oral, anal, and phallic components, in a manner
reminiscent of Freud's postulated stages of human
psychosexual development. Unequivocal sex differences
exist at early ages. Significantly more pelvic thrusting
is demonstrated by young males. Infantile female monkeys
rarely exhibit male-type behavior, and males rarely
exhibit responses of females. "Inappropriate" sexual
posturing is seen in young monkeys prior to effective
adult-type genital approximation. This is easily demonstrated by photographs taken during monkey "play periods."
Evidence is presented that a biological power exists
which underlies a monkey's reproductive ends, independent
of "training." These findings may bear significance with
respect to human sexuality.

223. Harlow, H.F., Harlow, M.K., Rueping, R.R., and Mason, W.A.
Performance of infant rhesus monkeys on discrimination learning,
delayed response, and discrimination learning set. Journal of
Comparative and Physiological Psychology, 1960, 53(2), 113-121.

    The performance of 65 infant rhesus monkeys on their
first object-discrimination problem was tested, and
comparisons were made among groups ranging from 60 to
360 days of age at the time of initial training. Mean
errors before problem solution were 31 for the 60-day
group and 7.5 for the 360-day group. The curve of
improvement for all groups was negatively accelerated.
The infant's performance was strongly influenced by
position and object preferences, and 12 infants, at least
1 from each group, attained the learning criterion without
making any errors. The performance of 55 of the above
subjects on 600 successive six-trial discrimination
problems and of 8 subjects on 400 problems was measured
after completion of the original discrimination problem.
At the end of 600 L-S problems Trial 2 performance for
Groups 60 and 90 was approximately 60%, for Group 120
about 70% and over 80% for Group 150. Group 360 was
much superior, approaching 80% correct Trial 2 responses
during the second 100 problems. The learning of zero-
second and 5-second delayed responses and subsequent
generalization to delays of 5, 10, 20, and 40 seconds

were measured in four groups of ten infants each, and
their data were compared with a group of ten adults.
There was clear evidence that performance improved as
a function of age and that an asymptote of performance
might be attained or approached shortly after eight
months of age. Although adult monkeys did not surpass
the two older infant groups in the generalized delayed-
response tests, they attained high level performance
on the zero and 5-second delays more rapidly.

224. Harlow, H.F., Harlow, M.K., Schiltz, K.A., & Mohr, D.J.
The effect of early adverse and enriched environments on the
learning ability of rhesus monkeys. In L.E. Jarrad (Ed.),
Cogitive processes of nonhuman primates. New York: Academic Press,
1971, 121-148.

Forty-five laboratory-raised rhesus monkeys (Macaca mulatta)
served as subjects and controls for this experiment.
Twelve animals were denied all social contact during
infancy, 12 were raised in socially enriched environments,
and 21 were control subjects. The monkeys ranged in age
from less than a year to over 2 years at the initiation
of testing, however the age ranges of the experimental
and selected control groups were comparable in that
representative experimental and control groups were tested
at each age level. All animals were examined in Wisconsin
General Test Apparatus (WGTA) with the following problems:
discrimination, discrimination learning set, delayed
response task, multiple delayed response task, and oddity
learning set. The results of this research seem to
indicate that previous social experience has a minimal
effect on learning ability in nonhuman primates.

225. Harlow, H.F., Joslyn, W.D., Senko, M.K., & Dopp, A.
Behavioral aspects of reproduction in primates. Journal of
Animal Science, 1966, 25, 49-65.

The present paper summarizes the development of reproductive
behaviors in nonhuman primates, particularly the common
rhesus macaque (Macaca mulatta). The development of
reproductive behavior is described in terms of five
heterosexual affectional stages: the reflex stage, the
infantile stage, the juvenile stage, the adolescent
stage, and the adult heterosexual stage. Although the
five stages are discretely described, it is recognized

that the development of reproductive behavior is a
continuous process in which originally simple and
relatively discrete reflexes are coordinated with and
complemented by more complex postural adjustments.
Throughout this discussion emphasis has been given to
the maturational components of these increasingly complex
patterns, even though it is recognized that learning
plays an extremely important role throughout the entire
reproductive development sequence.

226.   Harlow, H.F., Schlitz, K.A., Blomquist, A.J., & Thompson, C.I.
Effects of combined frontal and temporal lesions on learned
behaviors in rhesus monkeys.   Proceedings of the National Academy
of Science, 1970, 66(2), 577-582.

   Delayed response ability, and to a lesser extent visual
   discrimination performance, is seriously impaired by
   extensive bilateral damage to the frontal lobes.
   Reciprocal anatomical connections between the frontal
   and temporal lobes suggested that massive lesions in
   both lobes might produce an impairment more complete than
   that resulting from frontal lobectomy alone. Five monkeys
   were given combined bilateral frontal and anterior-
   temporal lesions, and were found to be inferior to
   both frontal lobectomized monkeys and to unoperated
   controls on the object discrimination task. The combined
   lesions did not increase the deficit on delayed response
   over that obtained after only bilateral frontal lobectomy.
   Results indicate that the anterior-temporal neocortex
   is involved in the mediation of visual discrimination
   ability.

227.   Harlow, H.F., Thompson, C.I., Blomquist, A.J., & Schlitz, K.A.
Learning in rhesus monkeys after varying amounts of prefrontal
lobe destruction during infancy and adolescence. Brain Research,
1970, 18, 343-353.

   Monkeys 5 and 24 months old were subjected to prefrontal
   topectomy or lobectomy, and were compared with like-aged
   controls on object discrimination, learning set, and
   oddity learning set. Size of lesion was the variable
   of primary importance in determining the extent of
   intellectual deficit. Total lobectomy produced significnat
   performance decrements on the object discrimination,
   delayed response, and object discrimination learning

set tasks. Differences between topectomized and control
monkeys did not reach statistical significance on any
of the tests utilized. There was little or no evidence
from the present study that surgery at 5 months spared
any intellectual functions that were lost after surgery
at 24 months. Older monkeys were superior to younger
monkeys on object discrimination, and to some extent
on oddity learning set. Younger monkeys were superior
to older monkeys on delayed response, but poor performance
by the older control group suggests that this result
may have been merely an age effect, rather than any
compensation of function occurring after early surgery.
Sparing of delayed response ability in monkeys operated
during infancy is a concept supported by data from other
sources, but the present study emphasizes the need for
including older control groups in future work in this
area.

228. Harlow, H.F., Blomquist, A.J., Thompson,C.I., Schiltz, K.A.,
& Harlow, M.K. Effects of induction, age and size of frontal lobe
lesions on learning in rhesus monkeys.   In R.L. Isaacson (Ed.),
The neuropsychology of development. New York: John Wiley and
Sons, 1968, 79-120.

This research examines the effect of lesions produced
in the frontal association cortex of rhesus monkeys
(Macaca mulatta) at ages ranging from 5 days to 2 years.
Delayed response was the measure chosen as most likely
to reveal lesion age effects. The strategy was to compare
the effects of lesions produced before ability to solve
delayed response has matured, at ages when the ability
is increasing in normal animals, and at an age when ability
is presumably complete or nearly complete. Discrimination
learning and learning set were chosen as measures likely
to show maximal sparing even though learning set is a
more difficult learning task than delayed response, as
indicated by ages at which infants can begin to master
the two problems and rate of improvement with increasing
age.

229. Haude, R.H. & Ray, O.S. Visual exploration in monkeys as a
function of visual incentive duration and sensory deprivation.
Journal of Comparative and Physiological Psychology, 1967, 64,
332-336.

Twenty-two rhesus monkeys (Macaca mulatta) were tested in two
experiments designed to quantify more precisely the
influence on visual exploratory behavior of 2 performance
variables, incentive magnitude and drive level. In
Experiment 1, which tested the effect of different rates
of changing color slides which served as visual incentives,
frequency of viewing and cumulative viewing time increased
as exposure duration of individual slides decreased.
In Experiment 2, no evidence of an effect of sensory
deprivation on visual exploration of color slide incentives
was found for any of the deprivation periods tested
(0, 2, 4, or 8 hr.).

230   Haude, R.H., Graber, J.G., & Farres, A.G.   Visual observing
by rhesus monkeys: Some relationships with social dominance rank.
Animal Learning and Behavior, 1976, 4(2), 163-166.

Five rhesus monkeys were tested in a visual exploration
situation to determine whether the mean frequency or
mean duration of visual observing were systematically
related to the dominance status of the observing animal.
A dominance hierarchy among the five subjects was first
determined by means of a competitive food-getting task.
Following dominance testing, visual exploration testing
was begun.   All subjects were permitted to observe all
other subjects in a round-robin pairing system involving
two animals at a time.   In each pairing, one animal served
as the experimental subject (observer); the other as
the stimulus (visual incentive).   A highly significant
linear effect of dominance was found in regard to
duration of observing.   Subjects high in the dominance
hierarchy observed for significantly shorter durations
than low-dominant subjects.   Significant effects of
dominance on the frequency of observing were also found,
with animals intermediate in the hierarchy viewing more
frequently than animals at either extreme.   The data
were interpreted in terms of the arousal and reduction
of fear as a function of dominance and also through the
notion of dominance distance.

231.   Hauty, G.T., & Yellin, A.M.   Activity cycles of the monkey
(Macaca mulatta) under different environmental conditions.   Journal
of Interdisciplinary Cycle Research, 1970, 1(2), 181-191.

Eight monkeys were subjected successively to four different
schedules of daily feeding and exposure to the experimenter for the purpose of appraising the effects of
these factors upon the form and amplitude of their daily
cycle of activity. The subjects were confined in
relatively unrestrained environments and committed to
a 12:12 L:D ratio. Each of the four schedules produced
a uniquely different curve of activity and the factor
of predominant influence appeared to be exposure of
the subjects to the experimenter. One finding, common
to all of the experimental schedules, consisted of a
high degree of photoperiodic dependency. To explore
the limits of such dependency, four monkeys were subjected
to three successive L:D schedules (12:12, 6:18, and
18:6). Essentially, the same degree of dependency
manifested under the 12:12 schedule was demonstrated
under the 18:6 schedule in that the onset and cessation
of activity closely followed the onset and termination
of the period of illumnation. A considerably lesser
degree of dependency occurred under the 6:18 schedule.
Here, the onset of activity preceded the onset of
illumination by 3-4 hours indicating a strong need for
a minimum period of activity of about nine hours.
Differences in habituation time were also observed.
Stable daily curves were obtained within 2-3 days following
the shift to the 6:18 schedule, 2-3 weeks following the
shift to the 18:6 schedule, and 2-3 days following
the shift back to the 12:12 schedule.

232. Hayes, C. The ape in our house. New York: Harper and
Brothers Publishers, 1951.

This book describes attempts made to raise a chimpanzee
as a human child. Data were obtained on the animal's
behavior from the age of three days to six and a half
years, and direct comparisons were made to human children
of the same age whenever appropriate. A serious effort
was made to teach her to speak, but after great labor
on the part of both Vicki and her "parents", the conclusion was reached that speech was extremely difficult
for a chimpanzee to master. She did, however, engage
in substantial two-way communication using gestures and
intention movements. Although Vicki's number-handling
skills remained that of a three-and-a-half-year-old
human child, her sorting abilities showed a high degree
of skill.

233.   Hayes, K.J. & Hayes, C.   The intellectual development of a
home raised chimpanzee.   <u>Proceedings of the American Philosophical
Society</u>, 1951, <u>95</u>(2), 105-109.

        This article deals with the intellectual development
        of a young chimpanzee whose background and experience is
        very much like that of a human child.   An extensive
        diary was kept of the chimpanzee's activities including:
        eye hand coordination, vocalizations, and verbalizing
        simple one syllable words.

234.   Hayes, K.J., & Nissen, C.H.   Higher mental functions of
home-raised chimpanzees.   In A.M. Schrier and F. Stollnitz (Eds.),
<u>Behavior of nonhuman primates</u>.   New York: Academic Press, 1971.

        This chapter reports on a battery of studies involving
        higher mental functions in a home-raised chimpanzee.
        Adopted into a home shortly after her birth, Viki was
        exposed to the standard toys, games, and household
        equipment, the family life and social experiences that
        form the environment of the average American preschool
        child.   Results indicate that to the extent that she
        was tested by special nonverbal techniques, Viki was
        found capable of handling at least rudimentary stages
        of certain "higher mental functions".

235.   Heath, R.G.   Electroencephalographic studies in isolation-
raised monkeys with behavioral impairment.   <u>Diseases of the
Nervous System</u>, 1972, <u>33</u> (3),   157-163.

        Deep and surface electrodes were implanted into the brains
        of a group of isolation raised monkeys with behavioral
        impairment and a control group of feral-raised monkeys
        with conventional behavior.   Recordings were obtained
        from both sensory relay nuclei and from limbic system
        structures implicated in emotional expression.   The
        severely disturbed isolation-raised monkeys not only
        showed electroencephalographic abnormalities from
        limbic system structures, but also from sensory relay
        nuclei for vestibular and proprioceptive function in the
        cerebellum and in the somatosensory thalamus.   The
        clinical evidence of a relationship between perception
        and emotion is substantiated by the physiologic
        demonstration in the isolation-raised monkeys.

236.   Heath, R.G.   Maternal-social deprivation and abnormal brain
development: Disorders of emotional and social behavior.   In
J. W. Prescott, M.S. Read, and D.B. Coursin (Eds.), Brain function
and malnutrition: Neuropsychological methods of assessment.   New
York: John Wiley & Sons, 1975,   295-310.

    This article presents data which pertains to cross-
    interpretations between brain mechanisms in rhesus monkeys
    (M. mulatta) and the animal's observed behavior.
    Preliminary data obtained from a small group of isolation-
    reared rhesus monkeys are the first evidence that aberrant
    electrophysiologic activity occurs in deep cerebellar
    nuclei, as well as in other deep brain structures--most
    pronounced in the limic system--in association with
    severely disturbed behavior resulting from maternal-
    social deprivation.   Further, a strong functional
    relationship between the cerebellum and limbic system
    structures has been demonstrated with use of electro-
    physiologic techniques.   Finally, the possibility of
    impaired brain function resulting from conditions of
    early social development that may co-exist with conditions
    of malnutrition during early development has been demon-
    strated and appropriate cautions must be raised concerning
    multiple interactions with malnutrition that may lead
    to an overstatement of the effects of malnutrition upon
    brain development and behavior.

237.   Held, R., & Bauer, J.A., Jr.   Visually guided reaching in
infant monkeys after restricted rearing.   Science, 1967, 155,
718-720.

    Infant macaques were reared from birth in an apparatus
    which precluded sight of their body parts.   At 35 days
    postpartum one hand was exposed to view.   Visual fixation
    of this hand was insistent and prolonged; visually
    guided reaching was poor, but it improved during ten
    succeeding hours of exposure.   Little concomitant
    improvement occurred in the reaching of the unexposed
    hand.

238.   Hendrickson, A., & Boothe, R.   Morphology of the retina and
dorsal lateral geniculate nucleus in dark-reared monkeys (Macaca
nemestrina).   Vision Research, 1976, 16, 517-521.

    Nine infant monkeys were reared in continuous darkness

from 2 weeks to 1.3 and 6 months of age. One monkey
was dark-reared from 3 to 7 months after birth. Light
microscopic morphological studies of retina and dorsal
lateral geniculate nucleus (dLGN) were done on animals
sacrificed immediately after emerging from darkness
and others that were tested behaviorally before sacrifice.
Neither retina nor dLGN showed any obvious changes in
cell number size or staining characeristics when
compared to light-reared, age matched controls.
Autoradiographic tracing of labeled retinal ganglion
cell synaptic terminals indicated a normal distribution
for dark-reared animals.

239. Hill, S.D., McCormack, S.A., & Mason, W.A. Effects of
artificial mothers and visual experience on adrenal responsiveness
of infant monkeys. Developmental Psychobiology, 1973, 6(5),
421-429.

Rhesus monkeys, maternally separated at birth, were individually housed with a simple artificial mother in cages
that were either completely enclosed (Enclosed Group)
or had a clear plastic front that faced the general
nursery environment (Visual Group). Animals were removed
from the living cage and placed for 1 hr. in an enclosed
carrying cage, either alone or with the artificial mother,
following which blood samples were taken and analyzed
for plasma cortisol levels. This procedure was repeated
at monthly intervals during the first 7 months of life
and at 9 and 12 months. In both groups plasma cortisol
levels were higher when the animals were alone than
when the social surrogate was present. Rearing conditions
were also effective, as reflected in longitudinal and
diurnal contrasts between groups: cortisol levels were
consistently high in the Visual Group from the 1st
week of life, whereas the developmental pattern was
curvilinear for the Enclosed Group. Significant diurnal
variation in cortisol level was demonstrated only in the
Visual Group.

240. Hill, S.D., Bundy, R.A., Gallup, G.G. Jr., & McClure, M.K.
Responsiveness of young nursery reared chimpanzees to mirrors.
VII Louisiana Academy of Sciences, 1970, 33, 77-82.

This study presents data which indicate that two-year-
old chimpanzees are capable of self-recognition. Although

complete access to peers appears to facilitate this
development, visual access to peers may not be sufficient
for the development of self-recognition.

241. Hinde, R.A. Mother-infant interaction in rhesus monkeys and
the consequences of maternal deprivation. In R.P. Michael (Ed.),
Endocrinology and human behavior. London: Oxford University Press,
1968, 3-11.

This paper is concerned with three aspects of mother-infant
interactions in rhesus monkeys (Macaca mulatta): (1) the
nature of changes in mother-infant interaction with age,
and of differences between mother-infant pairs; (2) the
influence of social companions on mother-infant interaction; and (3) the effect of a period of maternal
deprivation. Data is based on studies of small captive
groups, each consisting of a male, three or four females,
and their young, and were collected on check sheets
divided into half-minute periods.

242. Hinde, R.A. Analyzing the roles of the partners in a behavioral
interaction--mother-infant relations in rhesus macaques. Annals
of the New York Academy of Science, 1969, 159, 651-667.

In the paper, a method for teasing out the roles of two
individuals in a changing relationship is tested against
data concerned with various aspects of mother-infant
interaction in rhesus monkeys (Macaca mulatta).

243. Hinde, R.A. Influence of social companions and of temporary
separation on mother-infant relations in rhesus monkeys. In B.M.
Foss (Ed.), Determinants of infant behavior (Vol. 4) London:
Methuen, 1969, 37-40.

Data from two studies which examined the influence of
social companions and temporary separation on mother-
infant relations in rhesus monkeys (Macaca mulatta) are
presented. The first study found clear differences in
the mother-infant interactions between animals raised
in isolation and those reared in small groups. The
results of the second study show that some rhesus
infants are adversely affected by a 6-day period of
maternal deprivation. The severity of the effects

vary with the nature of the preseparation relationship,
but they may persist for some weeks.

244.   Hinde, R.A.   Development of social behavior.   In   A.M.
Schrier and F. Stollnitz (Eds.), Behavior of nonhuman primates
(Vol. III).   New York: Academic Press, 1971, 1-68.

It has been shown that the behavioral development of
rhesus monkeys (M. mulatta) may be grossly distorted
if they are reared without companions. Investigators have
also described the relationships formed when the infant
is allowed access to different categories of social
companions, and have specified many of the factors that
control the interactions involved. The present chapter
reviews work that extends these studies in three ways.
First, current knowledge of the social development of
rhesus infants is compared with data on other species.
Second, attention is focused more on the nature of the
interactions between the infant and its various social
companions, rather than on the long-term consequences
of those interactions. Third, the complexity of the
social environment is stressed.

245.   Hinde, R.A.   Some problems in the study of the development of
social behavior.   In   E. Tobach, L.R. Aronson and E. Shaw (Eds.),
The biopsychology of development.   New York: Academic Press, 1971,
411-432.

This chapter discusses some problems in the study of social
development. Because their social relationships tend to
be more complex than those of other species, attention is
concentrated mainly but not entirely on subhuman primates.

246.   Hinde, R.A.   Mother-infant separation in rhesus monkeys.
Journal of Psychosomatic Research, 1972, 16, 227-228.

Data from a series of experiments on mother-infant
separations in rhesus monkeys (M. mulatta) are presented.
Factors which may affect the severity of effects on the
infant are: (1) whether testing takes place in a novel
or familiar environment; (2) length of separation; (3)
the nature of the mother-infant relationship; and (4)
whether infant or mother is removed from home cage
during separation.

247. Hinde, R.A. Mother/infant relations in rhesus monkeys. In
N.F. White (Ed.), Ethology and psychiatry. Toronto: University
of Toronto Press, 1974.

     This paper discusses mother-infant separation in rhesus
     monkeys (M. mulatta) and the relevance to humans of
     studies of subhuman species. The author suggests that
     the symptoms shown after a separation experience are
     very similar in monkeys and humans. The only major
     difference appears to be that the phase of detachment,
     or rejection of the mother by the infant after the
     separation experience, is not present in monkeys--at
     least not to the same extent as in humans.

248. Hinde, R.A., & Davies, L.M. Changes in mother-infant
relationship after separation in rhesus monkeys. Nature, 1972,
239, 41-42.

     If separated from their mothers for a few days, infant
     rhesus monkeys and other macaques show increased distress
     calling, and reduced locomotor activity and other symptoms
     of behavioral depression. The symptoms may persist
     for weeks after reunion with the mother, and infants
     which have had such a separation experience differ from
     non-separated controls in their responses to strange
     objects months and even years later. The trauma of the
     separation experience itself may not be the only or
     even the major factor in producing these effects. The
     disturbance to the mother-infant relationship which
     results from the separation seems to be of considerable
     importance.

249. Hinde, R.A., & Davies, L. Removing infant rhesus from
mother for 13 days compared with removing mother from infant.
Journal of Child Psychology and Psychiatry, 1972, 13, 227-237.

     In a previous experiment, 6 rhesus monkey mothers were
     removed from their infants for 13 days, the infants
     being left in their familiar environment. Here the
     procedure was reversed--5 infants were removed to an
     isolation cage in a strange room, the mothers staying
     in their familiar environment. Contrary to expectation
     these infant-removed infants were less disturbed than
     the mother-removed infants. This difference was
     associated with, and was probably due to, a more positive

role by the mothers of the infant-removed group in mother-
infant interaction after reunion. It is suggested that
the effects of a separation experience may be mediated
in part by the disturbance to the mother-infant relationship which results.

250. Hinde, R.A., & McGinnis, L. Some factors influencing the
effects of temporary mother-infant separation: Some experiments
with rhesus monkeys. Psychological Medicine, 1977, 7, 197-212.

Some experiments, reported in detail elsewhere, on the
effects of mother-infant separation in rhesus monkeys
are here reviewed and compared. They involved 4 groups:
one in which mothers were removed for 13 days leaving the
infant in the social group; one in which infants were
removed; one in which mothers and infants were removed and
separated; and one in which mothers and infants were
removed but not separated. The nature of the separation
experience had a profound effect on the infant's response:
infants left in a familiar environment while their mothers
were removed showed marked but brief 'protest' and
then profound 'despair', whilst infants removed to a
strange cage showed more prolonged 'protest'. A major
factor determining the effects of the separation
experience in the weeks following reunion is the degree
to which the mother-infant relationship has been disturbed
by it. The multiplicity of factors affecting the outcome of a separation experience is discussed.

251. Hinde, R.A., & Spencer-Booth, Y. The behavior of socially
living rhesus monkeys in their first two and one-half years.
Animal Behavior, 1967a, 25, 169-196.

The development of behavior in eight rhesus monkey (M.
mulatta) infants living with their mothers in small
social groups was studied for the first two and one-
half years of life. Eight others were observed for
their first 24 to 30 weeks. The changes in various
aspects of behavior and infant-mother relations with
age were assessed, and the results indicate that the
mother plays a large part in the increasing independence
of the infant with age. At any particular age both
infant and maternal characteristics are important in
determining the time spent off the mother and related
measures. Those infants who sought the proximity of

their mothers more often were also those which were
rejected more often, spent more time on their mothers,
but more of their time off at a distance from her. Some
factors determining individual differences are discussed.

252. Hinde, R.A., & Spencer-Booth, Y. The effects of social
companions on mother-infant relations in rhesus monkeys. In
D. Morris (Ed.), <u>Primate ethology</u>. London: Weidenfeld & Nicolson,
1967b.

Data are presented which demonstrate differences in the
nature of the mother-infant interactions between isolated
mother-infant pairs and those pairs which live in groups.
Differences in mother-infant relations are ascribed to
two main environmental differences. First, in the
absence of aunts the isolated mothers were less restrictive
and their infants ranged to a distance from their mothers
more freely. Second, in the absence of play-companions,
during the second six months, the isolated infants returned
more often to the only other animal in their pen, their
mother, and thus spent shorter periods at a distance
from her.

253. Hinde, R.A., & Spencer-Booth, Y. Individual differences in
the responses of rhesus monkeys to a period of separation from
their mothers. <u>Journal of Child Psychology and Psychiatry</u>, 1970,
<u>11</u>, 159-176.

(1) Individual differences in the responses of 16 rhesus
monkey infants to a 6-day period of separation from
their mothers were examined. (2) On the day the mother
returned, a correlational matrix of various measures
of mother-infant interaction indicated that individual
differences between mother-infant pairs depended rather
more on differences between mothers (as opposed to
differences between infants) than before separation.
There was a gradual return to the pre-separation condition
during the next few weeks. (3) Measures of the infant's
behavior and of mother-infant interaction on the day
the mother was returned were not significantly correlated
with their preseparation values. (4) During the next
week, and to a lesser extent subsequently, the rejection/
acceptance interactional behavior between mother and
infant, and the relative role of the infant in maintaining
proximity, could be predicted from their preseparation
values. The times the infant spent off and at a distance

from the mother could not. (5) Those infants showing
greatest disturbance (as indicated by a "distress index")
after separation tended to be those which had the highest
frequency of rejections, and played the greatest role
in maintaining proximity to their mothers, both before
separation and contemporaneously.

254.   Hinde, R.A. & Spencer-Booth, Y.   Effects of brief separation
from mother on rhesus monkeys.   Science, 1971, 173(3992), 111-118.

Presents data on the development of mother-infant relations
in rhesus monkeys (Macaca mulatta) by examining the effects
of brief separations. Ss were 16 infants observed during
the 1st 6 mos. of life, 8 of whom were observed beyond
that period. Methods for separating the relative roles
of mother and infant are described. It was found that
when a mother was separated from her infant for a few
days, the infant would call a great deal initially and
would then show reduced locomotor and play activity.
Such symptoms could last for a mo. after the mother's
return. Tests given 6 mo. and 2 yr. later strongly suggest
that differences between infants that have had such a
separation experience and those that have not are persistent.
The relevance of these findings for human behavior is
discussed.

255.   Hinde, R.A., & Spencer-Booth, Y.   Towards understanding
individual differences in rhesus mother-infant interaction.   Animal
Behavior, 1971, 19, 165-173.

The rank orderings of an individual rhesus infant on
various measures of mother-infant interaction show
considerable stability over the first 24 weeks of
life. Some of the individual differences could be related
to the sex of the infant, or to the parity or dominance
status of the mother, but these status variables did
not appear to be of major importance. Matrices of
correlations between measures of mother-infant interaction
in successive age spans permitted some separation of the
roles of mothers and infants in causing the individual
differences observed. Early on, and especially in weeks
7 to 12, inter-mother differences were of prime importance,
but later inter-infant differences came also to play a
large role.

256. Hinde, R.A., Leighton-Shapiro, M.E., & McGinnis, L. Effects
of various types of separation experience on rhesus monkeys 5
months later. Journal of Child Psychology and Psychaitry, 1978,
19, 199-211.

   Thirty-week old infant rhesus monkeys, living with their
   mothers in small groups, were subjected to four types
   of treatment. Either the mothers were removed for 13
   days, the infants remaining in the colony (M/R), or the
   infants were removed (I/R), or both mother and infant
   were removed and separated from each other (M-I/R), or
   both were removed for 13 days but kept together [(MI)/R].
   Control monkeys were left in the home groups. At 1
   year data were collected in the home pens, and after mother
   and infant had been removed to test cages. The data
   support the view that a period of separation between
   mother and infant can, but need not, have long-term
   sequelae. The M/R infants, most affected in the weeks
   following reunion, were also the most affected at 1 year.
   Group differences in behavior in the test cages suggested
   that adverse conditioning to the separation environment
   played an important role, though this did not operate
   in precisely the same way on mothers and infants. It
   is emphasized that the consequences of a separation
   experience are complex.

257. Hinde, R.A., Spencer-Booth, Y., & Bruce, M. Effects of
6-day maternal deprivation on rhesus monkey infants. Nature,
1966, 210, 1021-1023.

   The present study presents data on mother-infant interactions in rhesus monkeys (Macaca mulatta) as a function
   of maternal deprivation. Results indicate that infants
   are adversely affected by a 6-day removal of their
   mothers, that the severity of the effect varies with the
   nature of the pre-separation relationships, and that
   these effects may persist for some weeks even after the
   mother is returned.

258. Hines, M. The development and regression of reflexes,
postures, and progression in the young macaque. Contributions
to Embryology, 1942, 30(196), 153-209.

   This paper examines the reflexes, posture, progression,
   and use of somatic musculature which characterizes the

infant macaque (Macaca mulatta) during the last month
of gestation and the first year of life. Data support
the general conclusion that the development and regression
through which these phenomena pass are an expression
of differential maturation of the central nervous system,
and in particular of the cortex cerebri.

259.   Hinkle, D.K. & Session, H.L.   A method for hand rearing of
Samiri sciureus.   Laboratory of Animal Science, 1972, 22(2)
207-209.

A method for hand rearing newborn and very young New
World monkeys (Saimiri sciureus) was described. There
has been an increased demand for this type of animal
in toxicological investigations in which the experimental
design requires separation of the mother from the neonate.
The infants were housed in incubators for the first 8
weeks of life and then transferred to individual stainless
steel cages. Their diet consisted of liquid formula
made from a combination of several commercial human
infant nutriments. This report includes mean values for
increase in body weight and formula consumption, as
well as some clinical observations during the first
12 weeks of life.

260.   Howells, J.G.   Of monkeys and men.   Acta Paedopsychiatrica,
1972, 38, 285-287.

This paper discusses the distinction between separation
and deprivation in nonhuman primates. The author maintains that it is not separation of the infant from the
mother per se that is damaging to the offspring, but
the consequent deprivation. Implications for separation
procedures in the human field are considered.

261.   Jacobsen, C.F., Jacobsen, M.M., & Yoshioka, J.G. Development
of an infant chimpanzee during her first year. Comparative
Psychology Monograph, 1932, 9(1), 94.

This monograph presents the life history of a chimpanzee
born of parents of known psychobiological characteristics.
The subject of description was followed from conception,
through a gestational period of nine lunar months, to

birth, and thereafter through the period of infancy.
The physical growth and physiological and behavioral
development of the animal are described.

262. Jensen, G.D. Reaction of monkey mothers to long-term
separation from their infants. Psychonomic Science, 1968, 11(5)
171, 172.

   Five pigtailed monkey mothers (Macaca nemestrina) were
   separated from their six-month-old infants. Mothers
   reacted partially with agitation. Eighteen days after
   separation several behavioral measures suggested depression.
   Two months after separation all measures returned to
   preseparation levels and showed no change after infants
   were reunited with their mothers.

263. Jensen, G.D. Effects of modificaiton of the social
environment on young monkey development. Biological Psychiatry,
1975, 10(6), 659.

   Experiments and observations of the effects of
   modifications of the social environment on the young
   primate's relationship with its mother and its social-
   sexual development have begun to delineate the large
   number of variables that determine and affect the attachment relationship and that are contingent upon it.
   Attachment can occur to a number of mother substitutes,
   including a dog. Early experience has special importance
   but young monkeys are surprisingly responsive and
   adaptable to substitute maternal stimuli. The quality
   of developing social behavior is dependent on both the
   quantity and quality of an attachment. We have yet
   to learn what are the critical component stimuli which
   make up an adequate early social environment for young
   monkeys and apes.

264. Jensen, G.D., & Bobbitt, R.A. Implications of primate
research for understanding infant development. In J. Hellmuth
(Ed.), The exceptional infant. Seattle: Special Child Publications,
1967, 515-542.

   This chapter describes a number of nonhuman primate
   studies completed and underway in various laboratories

which have challenging implications for theories of emotional
disturbance in human beings. For example, knowledge of
the importance of peer relationships may well have significant effects on the concept of primacy of the mother-
infant relationship in emotional health and disorders of
humans. Knowledge of the important role of peers may
have practical application in devising programs for
optimal institutional and foster parent care. The
primate studies on environmental enrichment are basically
relevant and supportive to current theories about the
fundamental importance of an adequate functional environment for optimal development of children in our society.

265. Jensen, G.D., & Bobbitt, R.A. Monkeying with the mother myth.
Psychology Today, 1968, 1, 41.

This article presents data generated from studies of the
mother-infant relationship of monkeys in both normal and
abnormal situations. The results of primate research
to date leads away from thinking in terms of the exclusive
importance of mother, peers, or the environment in early
life.

266. Jensen, G.D., & Tolman, C.W. Activity level of the mother
monkey (Macaca nemestrina) as affected by various conditions of
sensory access to the infant following separation. Animal Behavior,
1962, 10, 228-230.

The present investigation was directed toward discovering
whether the gross motor activity displayed by the mother
after separation from her infant was simply an all-or-
none effect of physical separation, or if the level
of activity would vary with different degrees of sensory
access to the infant following separation. In order
to show whether or not the effect was exclusively maternal,
a second experiment was performed duplicating the first
procedure, using non-parturient adult females. The
results indicate that the gross motor activity level
of the mother monkey is systematically affected by the
amount of sensory access to the infant as varied in
this experiment, whereas the activity of non-mothers
is affected to a lesser degree.

267. Jensen, G.D., & Tolman, C.W. Mother-infant relationship in
the monkey (Macaca nemistrina): The effects of brief separation
and mother-infant specificity. Journal of Comparative and
Physiological Psychology, 1962, 55(1), 131-136.

   A method was described by which the infant-directed
   behavior of mother monkeys (Macaca nemestrina) and the
   mother-seeking behavior of infants can be measured.
   Both these measures were taken to quantify the mother-
   infant tie. An experiment using this method was reported
   in which behavior was measured following brief separation.
   Observations were made of mother-own-infant and mother-
   other-infant combinations. The results seemed to
   warrant the following conclusions for mother-infant
   couples raised together until five to seven months of
   age: 1) Infant-directed behavior of mothers appears to
   be generally own-infant-specific, 2) Infants are not
   absolutely mother-specific, but they learn to be so
   after repeated separation and interaction with a strange
   mother, 3) Separation serves to increase the infant-
   directed behavior of the mother toward her own infant
   and the mother-seeking behavior of the infant for an
   initial period after they are reunited.

268. Jensen, G.D., Bobbitt, R.A., & Gordon, B.N. The development
of mutual independence in mother-infant pigtailed monkeys (Macaca
nemestrina). In   S.A. Altmann (Ed.), Social communication
among primates. Chicago: University of Chicago Press, 1967.

   This paper examines the roles played by the mother and the
   infant monkey (Macaca nemestrina) in the detachment
   process, and the ways in which differences in environment
   affect this process. Data suggest that environmental
   privation, without any maternal deprivation, retards the
   mutual independence or detachment process and the process
   of reattachment to other stimulus objects, and that this
   degree of early environmental privation critically affects
   the animals' later social performance.

269. Jensen, G.D., Bobbitt, R.A., & Gordon, B.N. Sex differences
in social interaction between infant monkeys and their mothers.
Recent Advances in Biological Psychiatry, 1967, 21, 283-293.

   This paper focuses on sex differences in mother-infant
   interaction in pigtails (Macaca nemestrina) developing

in two contrasting environments: a highly simplified or
"privation" one and a "rich" one. Particular measures
that suggest deviant behavior or some displacement in
the usual course of development for the mother or the
infant during the first 15 weeks of the infant's life
are reported.

270. Jensen, G.D., Bobbitt, R.A., & Gordon, B.N. Effects of
environment on the relationship between mother and infant pigtailed
monkeys (Macaca nemistrina). Journal of Comparative and Physiological
Psychology, 1968, 66(2), 259-263.

Eight mother-infant pairs of Macaca nemestrina were housed
in bare cages located in soundproof rooms and 4 pairs
in the laboratory in cages containing toys. Interactive
behavior was observed during the 1st 15 wk. of the
infants' lives. Pairs in the privation environment
spent more time in physical contact; infants of these
pairs oriented more behavior toward themselves and less
toward the environment, locomoted less, did less climbing,
and showed less differentiation in manipulation of
their mothers.

271. Jensen, G.D., Bobbitt, R.A., & Gordon, B.N. Studies of
mother-infant interactions in monkeys (Macaca nemestrina).
Proceedings of the 2nd International Congress of Primatology,
Atlanta, G.A., 1968, 1, 186-193 (Karger, Basel/New York, 1969)

In an analysis of mothers' hitting behavior directed at
their infants in the first 15 weeks of the infant's
lives, one mother was studied with successively-born
female infants raised in two different environments, one
poor and one rich in stimuli. Significant patterns of
behavior and sequences of patterns that characterized
this mother and her infants in each environment were
compared with larger groups of mother-infant pairs in
the two different rearing environments. In the more
normal environment, the mother's hitting was effective
in inducing the infant to leave. In the privation
environment, where infants characteristically did more
climbing-on, hitting was ineffective in instigating
independence.

272. Jensen, G.D., Bobbitt, R.A., & Gordon, B.N. Dominance testing
of infant pigtailed monkeys reared in different laboratory environments. Proceedings of the 3rd International Congress of Primatology,
Zurich, 1970, 3, 92-99 (Karger/Basel, 1971)

    The differential effects of 3 rearing environments on
    the dominance relationships of 12 mother-raised infant
    pigtailed monkeys were studied. Infants were paired
    and the behavioral interactions between members of each
    pair were observed in free-interaction and food competition
    settings. Consistent dominant-submissive patterns of
    interaction were found. Infants reared in more stimulating
    environments dominated those reared in poorer environments.
    The adverse effects of early deprivation were ameliorated
    by environmental enrichment at 4 months of age.

273. Jones, B.C., & Clark, D.L. Mother-infant separation in
squirrel monkeys living in a group. Developmental Psychobiology,
1973, 6(3), 259-269.

    Two female and 3 male infant squirrel monkeys were
    separated from their mothers at a mean age of 169 days.
    All had been group reared. The experiment entailed
    three 5 day periods: 1) preseparation; 2) separation; and
    3) reunion. During the experiment each infant was observed
    for 10 min. per day. Frequency and duration measures
    were recorded for activity play, locomotion, object
    manipulation, and disturbance. Social behaviors recorded
    were contact play, mixed play, competition play, sex
    play, affiliation, display, and aggression. Changes
    noted during separation were depression of activity play,
    contact play, mixed play, object manipulation, and sex
    play. Affiliative behaviors and locomotion increased
    during separation. After reunion, locomotion and
    affiliation returned to baseline levels while contact
    play, mixed play, and object manipulation remained
    depressed.

274. Jones, I.H. Ethology and psychiatry. Australian and New
Zealand Journal of Psychiatry, 1971, 5(4), 258-263.

    Some of the principles underlying the science of ethology
    are described. Analogies are drawn between ethological
    and psychiatric observations in the psychiatric areas
    of reactions to separation, bereavement, depression,

anxiety, sexual disorders and hysteria. It is suggested
that in other areas also, notably obsessional states and
schizophrenia, ethological concepts may be relevant.
The implications and limitations of these analogies are
briefly explored.

275.   Joslyn, W.D. Behavior of socially experienced juvenile
rhesus monkeys after eight months of late social isolation and
maternal-offspring relations and maternal separation in juvenile
rhesus monkeys. (Doctoral dissertation, University of Wisconsin,
1967), Dissertation Abstracts International, 1968, 28(3).

Six socially experienced juvenile rhesus monkeys (M. mulatta)
were deprived of social contact with other monkeys between
18 and 26 months of age. Upon emergence from isolation,
they immediately engaged in playful, sexual and affiliative
behaviors appropriate to their age. Another group of
six monkeys lived with their mothers between the ages
of 18 and 26 months after having been apart from their
mothers for a total of four months. Although the mother-
offspring relationship remained infantile, this prolonged
maternal indulgence did not inhibit social development
in terms of peer relations. In a follow-up study all
twelve of the juvenile monkeys used in the experiment
were returned to their mothers for one week at the age
of 30 months. All subjects resumed an infantile maternal
relationship even though they were almost sexually mature.

276.   Kaplan, J. The effects of separation and reunion on the
behavior of mother and infant squirrel monkeys. Developmental
Psychobiology, 1970, 3(1), 43-52.

Mother and infant squirrel monkeys that lived together
in a socially restricted environment were separated for
a period of seven days after the infants had become
relatively independent, and were then reunited. Behavioral
observations before, during, and after separation indicated
that 1) female infants became independent of their mothers
earlier than males, 2) neither mothers nor infants were
severely affected by separation, and 3) an increase in
attachment occurred following reunion only when the
mother had a limited history of maternal experience.
These results suggest that 1) certain characteristics
in the maternal behavior of the squirrel monkey facilitate
readjustment in mothers and infants following maternal

separation, 2) maternal experience can influence the
mother-infant relationship following a period of separation,
and 3) the independence of infants from their mothers
may be a function of both sex and species.

277.   Kaplan, J. Differences in the mother-infant relations of
squirrel monkeys housed in social and restricted environments.
Developmental Psychobiology, 1972, 5(1), 43-52.

Mother and infant squirrel monkeys were housed together
in either a group or a restricted environment until the
infants were approximately 22 weeks of age. Observations
of the mothers' and infants' behavior during this period
revealed clear differences between the rearing conditions.
Mothers in the socially restricted environment avoided
and punished their infants more and were generally less
protective than those in the group environment. However,
these differences did not appear to reflect differences
in maternal attitudes per se, but rather the extent to
which infants in the two environments engaged in certain
activities. Infants in the restricted environment
attempted to play with their mother more often and
remained closer to her as they became older.

278.   Kaplan, J. Growth and behavior of surrogate-reared squirrel
monkeys. Developmental Psychobiology, 1974, 7(1), 7-13.

Infant squirrel monkeys of the Colombian and Peruvian
varieties were reared on surrogates until they were
24 weeks old. The weights of the Colombian animals were
consistently greater than those of the Peruvian animals,
and the males of each type were heavier than the
respective females. The behavior of both groups was
similar in most respects and much like that reported
for surrogate-reared rhesus infants. However, the
Colombian infants did spend less time than the Peruvians
on the surrogate during the first 12 weeks, and none
of the animals exhibited anything resembling the stereotyped body-rocking that has been described for the
surrogate-reared rhesus.

279. Kaplan, J. Perceptual properties of attachment in surrogate-reared and mother-reared squirrel monkeys. In S. Chevalier-Skolinkoff and F.E. Poirer (Eds.), Primate bio-social development: Biological, social, and ecological determinants. New York: Garland Publishing Inc., 1977, 225-234.

This article discusses a series of studies undertaken
with both surrogate-raised and mother-raised infant
squirrel monkeys (Saimiri sciureus) which examined how
simple perceptual characteristics of the mother are
involved in the development of filial attachment.
Results show that squirrel monkey infants can recognize
and prefer their mothers when olfactory cues appear to
be the major source of information. Moreover, the
addition of behavioral cues and information from other
sensory modalities tends to enhance this recognition.

280. Kaplan, J. Some behavioral observations of surrogate- and mother-reared squirrel monkeys. In S. Chevalier-Skolnikoff and F.E. Poirier (Eds.), Primate bio-social development: Biological, social, and ecological determinants. New York: Garland Publishing Inc., 1977, 501-514.

The present chapter comprises some observations made by
the author (with others) of groups of infants reared with
artificial mother substitutes or with real mothers.
This research was undertaken to help clarify phylogenetic
distinctions among primates. Preliminary results suggest
that the squirrel monkey is less affected than the rhesus
with respect to at least one type of early rearing treat-
ment--that of being raised both without a real mother
and without peers for the first six months of life.

281. Kaplan, J., & Schusterman, R.J. Social preferences of mother and infant squirrel monkeys following different rearing conditions. Developmental Psychobiology, 1972, 5(1), 53-59.

Mother and infant squirrel monkeys that were previously
housed together in either a social or restricted environment were given the opportunity to select each other, an
unfamiliar female, an unfamiliar infant, or an empty
cage, in a free-choice situation. The time spent with
the different choices was used as an index of the animal's
preference. Five of the six socially reared infants
clearly preferred their mother to the other choices,

whereas only one (the youngest) of the six infants reared
alone with its mother showed a preference for her. Most
of the mothers in both groups did not show any preference
among the four choices, although two mothers from the
social group and two from the restricted group did spend
considerably more time with their own infant. The
difference between the two groups of infants is discussed
in terms of differences in the environmental characteristics and mother-infant relationship of the two rearing
environments.

282. Karrer, R., & Dekker, L. Preliminary observations on the
development of dominance in partial-isolation-reared monkeys (Macaca
irus). Bulletin of Psychonomic Society, 1973, 2(4), 225-228.

Macaca irus, reared without physical contact with other
monkeys for 6 of 12 months after birth, were deficient
in dominance-submissiveness behaviors. A relatively
short period of group experience at age 2.5 years resulted
in the development of dominance-submissiveness behavior.
Therefore, social experience later in development appears
to be effective in reversing some of the consequences
of early social deprivation.

283. Kaufman, I.C. Some biological considerations of parenthood.
In E.J. Anthony and T. Benedek (Eds.), Parenthood: Its psychology
and psychopathology. Boston: Little, Brown & Co., 1970, 3-55.

The biological aspects of parenthood are discussed from
an evolutionary perspective with an emphasis on nonhuman
primate research. Data from the laboratory of the author
and others are presented on mother-infant relations in
two closely related species of macaque, pigtails (Macaca
nemestrina) and bonnets (Macaca radiata), and the results
discussed.

284. Kaufman, I.C. Mother-infant separation in monkeys: An
experimental model. In   J. Scott and E. Senay (Eds.), Separation
and depression. Publication No. 94 of the American Association
for the advancement of Science, 1973, 33-52.

This paper discusses reactions to separation from, and
subsequent reunion with, mother in infant pigtail

(M. nemestrina) and bonnet (M. radiata) monkeys. Behavioral
differences between the two species are discussed.

285. Kaufman, I.C. The role of ontogeny in the establishment of
species-specific patterns. In J.J. Nurnberger (Ed.), Biological
and environmental determinants of early development. Baltimore:
Williams, and Wilkins, 1973.

    This paper discusses two species of macaques (M. nemestrina
    and M. radiata) for the purpose of illustrating the
    role of life experience in the establishment of behavior
    patterns that are species-specific. Inasmuch as monkeys
    are very social creatures, data is presented from the
    study of animals living in groups in order to assess,
    as much as possible, the full behavioral repertoire
    as it exists in natural troops with animals of both
    sexes and all ages. Two closely related species were
    studied in the hope that this greater base of comparative
    data might enhance the ability to make evolutionary
    generalizations and extrapolations.

286. Kaufman, I.C. Mother-infant relations in monkeys and humans:
A reply to Professor Hind. In N.F. White (Ed.), Ethology and
psychiatry. Toronto: University of Toronto Press, 1974, 47-58.

    In this chapter Dr. Kaufman reviews a paper by Dr. Hinde
    which concerns itself with mother-infant relations in
    nonhuman primates. Results from Kaufman's own research
    in this area are presented, and similarities and differences
    between the two are discussed.

287. Kaufman, I.C. On animal models. In R.J. Friedman and M.M.
Katz (Eds.), Psychology of depression: Contemporary theory and
research. Washington, D.C.: Winston-Wiley, 1974, 251-153.

    In this panel discussion, Dr. Kaufman makes the following
    points: 1) Underlying all depressive reactions is a
    basic organismic state, a state of helplessness; 2) This
    organismic state is based on a biological response system,
    present throughout much of the vertebrate order, which
    is available for dealing with unmanageable stress by
    favoring survival through the adaptive aspects of
    conservation-withdrawal response; 3) The state of

helplessness should be viewed as a psychobiological
entity in which motivational, cognitive, and affective
processes are interdependent variables; 4) If a state
of helplessness is the basic phenomenon of depression,
we need not postulate a primary role for aggression in
depression; and 5) The role of aggression in depression
is obscured by a conceptual confusion between the causes
and effects of behavior and also by a semantic and
conceptual confusion involving the relationships among
"aggression", "hostility", and "anger".

288.   Kaufman, I.C.   Learning what comes naturally: The role of
life experience in the establishment of species typical behavior.
Ethos, 1975, 3(2), 129-142.

This paper presents data from a long-term study of two
species of macaques, bonnets (M. radiata) and pigtails
(M. nemestrina) which suggest that in each species there
are predispositions to learn more easily certain things
from species-typical life experiences. With respect to
these two species, the conclusion drawn is more specific,
and states that the differences in social relations,
expecially maternal behavior, are the particular species-
typical life experiences that are crucial to the social
development of infants in each species as they grow
to maturity.

289.   Kaufman, I.C.   Developmental considerations of anxiety and
depression: Psychobiological studies in monkeys. In   T. Shapior
(Ed.), Psychoanalysis and contemporary science, Vol. 5. New York:
International Universities Press, 1977, 317-363.

This paper describes some of the author's studies of
monkeys in an effort to illuminate the understanding
of anxiety and depression, with particular emphasis on
two topics, the processes of adaptation and the functional
transition from the biological to the psychological
level of organization. From use of the experimental
paradigm of mother loss, data are presented to document
three main conclusions. First, the infant reacts to this
stress adaptively, using both inborn and acquired coping
mechanisms. Second, in the course of development the
infant's reactions become psychobiological rather than
purely biological as the inborn biological response
systems of flight-fight and conservation-withdrawal are

elaborated cognitively, motivationally, and affectively
into the organismic states of anxiety and depression
respectively. Third, depression is basically a state
of helplessness.

290. Kaufman, I.C., & Rosenblum, L.A. A behavioral taxonomy for
Macaca nemestrina and Macaca radiata: Based on longitudinal
observation of family groups in the laboratory. Primates, 1966,
7(2), 205-258.

This work presents a detailed blueprint of the structure
of behavior in two congeneric species of monkey, bonnet
(M. radiata) and pigtail (M. nemestrina). It is hoped
that the descriptions and illustrations will serve as
a useful guide to investigators interested in carrying
out research on the behavior of these and related species
by providing them with an initial focus on behaviors of
significance, their discriminable elements, and the
relevant contexts in which they fall. Data were recorded
systematically for frequency, duration and object of
each behavior, and in addition for sequences of behaviors
for given animals.

291. Kaufman, I.C, & Rosenblum, L.A. Depression in infant monkeys
separated from their mothers. Science, 1967, 155, 1030-1031.

The mothers of four pigtail (Macaca nemestrina) infants
living in a group were removed for 4 weeks. All infants
reacted initially with agitation. Three of the four
infants then became severely depressed. The depression
lasted about a week and was strikingly similar to the
"anaclitic depression" of human infants who lost their
mothers. When they were reunited, all four dyads showed
a marked and prolonged intensification of the mother-
infant relationship.

292. Kaufman, I.C., & Rosenblum, L.A. The reaction to separation
in infant monkeys: Anaclitic depression and conservation-withdrawal. Psychosomatic Medicine, 1967, 29(6), 648-675.

The reaction to removal of the mother was studied in 4
group-living pigtail monkey infants. All showed distress
with 3 progressing to a state of deep depression similar

to the anaclitic depression of human infants following
separation as described by Spitz. The only infant not
showing deep depression was the offspring of the dominant
female. The stages of reaction are seen as successive
efforts at adaptation based on available response systems,
evolved for their selective advantage or developed
ontogenetically, especially through dominance-hierarchical
regulatory influences. In this regard the reactions
have apparent survival value, in part through their
communicative significance. Monkey infants have a
greater chance of survival without a mother figure
than humans because of their greater locomotor ability
which appeared to initiate recovery from the depressed
state. The data support Engel's theory of two primitive
biological response systems for handling distress, each
with a mediating neural organization.

293. Kaufman, I.C., & Rosenblum, L.A. The waning of the mother-
infant bond in two species of macaque. In   B.M. Foss (Ed.),
Determinants of infant behavior (Vol. 4). London: Methuen & Co.,
1969a, 41-59.

This paper describes changes in the mother-infant relationship in two species of macaque, pigtail (M. nemestrina)
and bonnet (M. radiata), which were observed in laboratory
groups under identical conditions. The data presented
cover the first fifteen months of infant life before
siblings were born, and differences between the two
species with respect to rate of development of independence are discussed.

294. Kaufman, I.C., & Rosenblum, L.A. Effects of separation from
mother on the emotional behavior of infant monkeys. Annals of the
New York Academy of Science, 1969b, 159, 681-695.

This article discusses a series of mother-infant separations
in two species of monkeys, pigtails (Macaca nemestrina)
and bonnets (Macaca radiata), and compares the emotional
reactions of these animals with each other, with separated
rhesus infants (Macaca mulatta), and with children.

295. Kaufman, I.C., Stynes, A.J. Depression can be induced in a
bonnet macaque infant. Psychosomatic Medicine, 1978, 40(1), 71-75.

    Depression is a well-known reaction to separation in
    all species of macaques studied except the bonnet
    (M. radiata), in whom this reaction has never before
    been observed. We have produced the depressive response
    of postural collapse and social withdrawal in a bonnet
    infant by altering the social structure of the group
    in which it lived in a way calculated to interfere
    with the species-typical system of social support.

296. Kaufman, J.G. Behavior of infant rhesus monkeys and their
mothers in a free-ranging band. Zoologica, 1966, 51(1), 17-28.

    The behavior of infant rhesus monkeys and their mothers
    during the first three months after birth was studied
    in the free-ranging colony on Cayo Santiago. Comparison
    of the Cayo Santiago infants with those in laboratories
    and outdoor runs shows close agreement in the rates
    of development of most kinds of behavior. The exceptions
    were chiefly in social behavior, and were probably due to
    a combination of individual differences and the more
    complex social environment on Cayo Santiago.

297. Kawabe, S. Development of vocalization in monkeys reared in
isolation. International Journal of American Linguistics, 1969,
35(3), 260. (abstract)

    This report is concerned with the development of vocalization
    in Japanese infant monkeys raised in isolation. Sixty-
    three types of vocalization were classified into four
    vocal groups corresponding to the emotional situations
    in which the sounds were recorded: 1) fear, frustration;
    2) calls; 3) muttering sounds; and 4) sudden utterances
    with no apparent meaning.

298. Kellogg, W.N. Chimpanzees in experimental homes. The
Psychological Record, 1968, 18, 489-498.

    Between 1932 and 1968 there have been 6 major research
    projects by qualified investigators of rearing young
    chimpanzees in experimental homes. The object or purpose

of such experiments, although often misunderstood, is to
determine the genetic limitations of the animal when it
is given the enriched environment of the civilized
household. In this paper the behavior of the home-raised
chimpanzees in these studies is evaluated in relation
to that of human children in pre-school intelligence
tests, in toilet-training, in general adaptation to the
environment, and in communications.

299. Kellogg, W.N., & Kellogg, L.A.   The ape and the child.   New
York: McGraw-Hill Inc., 1933.

This book describes the home raising of a 7-month old
female chimpanzee for a period of nine months. The
researcher's infant son was used as a co-subject, and
comparisons were made of the abilities and progress of
the two infants. The chimpanzee, Gua, was precocious
in her early development, particularly locomotor skills,
and in the first half of the study was either equal
to the human infant or slightly superior as measured by
correct responses to verbal instructions. In the last
half of the experiment, however, the human infant's
linguistic competence overtook the chimp's, and he
surpassed her in comprehension.

300. Kennard, M.A. Cortical reorganization of motor function:
Studies on series of monkeys of various ages from infancy to
maturity. Archives of Neurology and Psychiatry, 1942, 48,
227-240.

This paper compares the effects of cortical ablations
on a series of 16 monkeys (Macaca mulatta) of varying
ages. Results indicate that the greatest capacity for
reorganization is during the first six months of the
animal's life. Additionally, the greatest loss of capacity
seems to occur during the end of the first year of life,
at the same time that spasticity begins to appear.
Finally, the recovery, previously shown to be due to
reorganization of function in the remaining areas of
the cortex, is slow and is maximal when ablations are
carried out seriatim and with long intervals between
extirpations. It is suggested that this is compatible
with the anatomic structure and growth of the cortex
and that the dendritic connections of the motor neurons
already present in the unexcised areas are reorganized
during the period of recovery of function.

301. Kerr, G.R., Chamove, A.S., & Harlow, H.F. Environmental
deprivation: Its effects on the growth of infant monkeys. The
Journal of Pediatrics, 1969, 75(5), 833-837.

   The nutritional intake of children with "deprivation
   dwarfism" has usually been reported to be normal, and
   in the absence of other etiologic explanations, their
   growth failure has been attributed to lack of environmental
   stimulation. Infant rhesus monkeys reared under conditions
   of total social isolation developed gross behavioral
   abnormalties, but grew at normal rates when fed an ad
   libitum diet. On the basis of these observations,
   inadequate nutrition would appear to be the most plausible
   explanation for the growth failure which may occur in
   children who are reared in deprived environments.

302. Kerr, G.R., Scheffler, G., & Waisman, H.A. Growth and
development of M. mulatta fed a standardized diet. Growth, 1969,
33, 185-199.

   Infant rhesus monkeys born after full term normal pregnancies
   were separated from their mothers within 4-8 hours and
   reared in individual cages under standard conditions.
   All animals were fed an ad libitum standardized diet at
   2-4 hour intervals for the first year of life. With the
   exception of a daily supplement of multiple vitamins,
   no other source of nutrition was provided. The daily
   dietary intake and body weight were recorded, and
   measurements of hematologic status, serum proteins,
   body length and head circumference were made at 15 to
   30 day intervals during the entire first year. All
   animals lost approximately 10% of their birth weight
   by 36 hours of age; this was regained by the end of the
   fourth postpartum day. The ad libitum diet intake, per
   Kg. body weight, increased during the first 4-6 weeks
   of life, then showed a progressive fall throughout the
   year; the ability to utilize dietary nutrients for
   physical growth showed a corresponding decrease. The
   rates of linear growth declined progressively for the
   first six months of life and thereafter remained at
   relatively constant levels. This experimental approach
   permits the evaluation of specific changes in this
   standardized diet with regard to their effect on the
   physical growth and functional development of a species
   whose growth processes are generally comparable to
   those of man.

303.   Kerr, G.R., Chamove, A.S., Harlow, H.F., & Waisman, H.A.
"Fetal PKU": The effect of maternal hyperphenylalaninemia during
pregnancy in the rhesus monkey (<u>Macaca mulatta</u>). <u>Pediatrics</u>, 1968,
<u>42</u>,(1), 27-36.

    Maternal hyperphenylalaninemia was induced by feeding
    excess L-phenylalanine to pregnant rhesus monkeys.
    Animals which received the phenylalanine supplement
    throughout the entire pregnancy gave birth to infants
    with low birth weight. Marked elevations of phenylalanine
    and tyrosine were consistently observed. The levels of
    phenylalanine and tyrosine in umbilical cord sera were
    greater than the corresponding maternal value in eight
    of nine pregnancies. Maternal and umbilical cord levels
    of serine were slightly elevated and maternal levels of
    3-CH3-histidine were reduced in all pregnancies. Values
    for the other amino acids, and the cord: maternal ratios
    for all amino acids, including phynylalanine and tyrosine,
    were within the control range. Infant monkeys born to
    mothers with hyperphenylalaninemia demonstrated a signif-
    icant reduction in learning behavior. This experiment
    indicates that not only is maternal hyperaminoacidemia
    reflected in the free amino acids of fetal blood, but
    an added insult is also produced by a normal placental
    process which functions to maintain higher levels of
    each of the free amino acids in the fetus than in the
    maternal organism.

304.   Kerr, G.R., Chamove, A.S., Harlow, H.F., & Waisman, H.A.
The development of infant monkeys fed low-phenylalanine diets.
<u>Pediatric Research</u>, 1969, <u>3</u>, 305-312.

    The nutritional adequacy of a commercial low-phenylalanine
    diet (Diet LF) has been investigated in infant rhesus
    monkeys. All animals were fed a control diet (Diet CD)
    during the first month of life. Thereafter, animals
    in Group A were fed LF until seventy-five days of age.
    This diet was then supplemented with 0.1g of L-phenylala-
    nine per kilogram of body weight until 105 days of age
    and with 0.2g of L-phenylalanine per kilogram from 105
    to 135 days of age. Animals in Group B were fed LF
    supplemented with an amount of L-phenylalanine equal to
    that contained in CD (0.087 g/100 ml) from 30 to 135
    days of age. For the remainder of the first year of life,
    all animals were again fed CD. Animals in Group A devel-
    oped growth failure, anemia, hypoproteinemia, dermatitis,
    edema, hypophenylalaninemia and elevated levels of

several other free amino acids in serum when fed LF.
The addition of supplements of L-phenylalanine corrected
hypoproteinemia, hypophenylalaninemia, and anemia,
but improvement of dermatitis and growth rates were
not seen until the aminals were again fed CD. Animals
in Group B developed dermatitis and elevated levels
of several of the free amino acids in serum but showed
no other biochemical or clinical evidence of phenylalanine deficiency. After one year of age, all animals
were evaluated for learning behavior. There was a
significant decrease in the learning ability of animals
in Group A, while that of animals in Group B was
comparable with that of control animals.

305. Kety, S.S., & Matthysse, S., (Eds.). Prospects for research
on schizophrenia. Neurosciences Research Program Bulletin, 1972,
10(4), 369-484.

This report, based on an NRP work session focuses on new
ideas and methods applicable to research on schizophrenia.
Drs. D.A. Hamburg, D. Ploog, and G.P. Sackett were among
participants who discussed some of the best animal models
available and how the study of these models can contribute
to an understanding of schizophrenia.

306. King, J.E., & King, P.A. Early behaviors in hand-reared
squirrel monkeys (Saimiri sciureus). Developmental Psychobioloby,
1970, 2(4), 251-256.

The development of behaviors during the first 50 days of
life was observed in four squirrel monkeys who had been
separated from their mothers shortly after birth. Behaviors observed included those related to nursing and maternal
contact as well as sensory development, locomotion, and
eating. Those behaviors closely related to maternal
contact and nursing were particularly strong and persistent
as compared to similar behaviors in rhesus monkeys, a
result consistent with the differences in the maternal
behaviors of these two species. Two of the monkeys who
survived to adulthood developed a stereotyped thumb
sucking and resultant malocclusions of the front deciduous
teeth, whereas the third surviving monkey displayed stereotyped in-place rocking.

307. Kling, A. & Dunne, K. Social-environmental factors affecting
behavior and plasma testosterone in normal and amygdala lesioned
M. speciosa. Primates, 1976, 17(1), 23-42.

    A social group of M. speciosa was observed in a laboratory
    enclosure and subsequently transferred to a 1/2 acre corral.
    Lesions of the baso-lateral amygdaloid nuclei in M. speciosa
    resulted in a disruption of affiliative behavior which
    did not recover over a three month period of continuous
    observation. Qualitative and quantitative measures
    indicated that no recovery in social isolation was
    evident. Social stress produced transient cohesiveness.
    Polymorphus sexual behavior increased iṅ both operates
    and normals while in a laboratory enclosure but was
    absent when placed in a 1/2 acre enclosure. Plasma
    testosterone levels were related to social rank in
    both males and females. Alterations in dominance in
    the post-operative period were followed by concomitant
    changes in testosterone levels.

308. Kling, A., & Green, P.C. Effects of neonatal amygdalectomy
in the maternally reared and maternally deprived macaque. Nature,
1967, 213, 742-743.

    This report discusses the effects of amygdala lesions in
    the monkey produced during the first week of life.
    The results suggest that (a) the effects of amygdalectomy
    are age-dependent and may require some degree of sexual
    maturation before becoming manifest; (b) the lack of
    influence of amygdalectomy on the maternal deprivation
    syndrome would indicate that this behavior is mediated
    by sub-cortical structures, is fixed rather early in
    life, and is not grossly influenced by higher brain
    systems; and (c) growth for the first year of life and
    appropriate infant-maternal transactions are not grossly
    affected by neonatal amygdalectomy.

309. Kling, A. & Tucker, T.J. Effects of combined lesions of
frontal granular cortex and caudate nucleus in the neonatal
monkey. Brain Research, 1967, 6, 428-439.

    Combined bilateral lesions of frontal granular cortex
    and caudate nucleus were produced in 6 monkeys between
    the 2nd and 56th postnatal day. When comparisons were
    made with 4 monkeys which had previously sustained

bilateral resections of frontal granular cortex alone at
comparable ages, it was found that the frontal-caudates,
in contrast to the frontals, displayed failure to thrive,
somato-motor and growth deficiencies, postoperative
seizures and later cognitive impairments. When tested at
the 7th postoperative month, 2 of the 3 surviving frontal-
caudates could not perform a delayed response task beyond
a minimal (0 sec) delay interval. It was concluded that
early reflexive and somato-motor function is principally
mediated by subcortical systems but that some degree of
representation also exists at cortical levels. In contrast
to neocortex, the caudate nucleus appears to be involved
in a neural system responsible for the retention of
delayed response capacity following early brain injury.

310. Kling, .A, & Tucker, T. Motor and behavioral development
after neodecortication in the neonatal monkey. Medical Primatology
1970. Proceedings of the 2nd Conference of exp. Med. Surg. Primates,
New York, 1969, pp. 376-393 (Karger, Basel 1971)

This report deals primarily with the long term somato-
motor function development of infant monkeys subjected
to extensive ablations of cerebral cortex, but sparing,
in some, localized anatomical areas. It is hoped that
long term survivals may shed light on the question of
the maturational stages at which certain cortical areas
are required for specific functions.

311. Kling, A., Dicks, D. & Gurowitz, E.M. Amygdalectomy and
social behavior in a caged group of vervets (C. aethiops).
Proceedings of the 2nd International Congress of Primatology,
Atlanta, Ga., 1968, 1, (Karger/Basel, New York: 1969)

The effects of amygdalectomy on social behavior in the
vervet were 1) a fall in rank to the most subordinate
level with respect to cage position, 2) lack of displacement behavior and fearless eating in spite of attack
by the normal animals, 3) absence of threat behavior
on the part of the operated animals, 4) the establishment
of a separate group of operates huddling together outside
the structure of the unoperated group, 5) lack of a
hierarchy within the operated group, and 6) an increase
in agonistic behavior by unoperated subjects. Effects
on individual behavior included 1) diminution of fear
toward man, 2) hyperorality and coprophagia, 3) prolonged

anorexia in adults, and 4) loss of vocal and postural
threat behavior.

312. Kling, A., Steinberg, D., Berman, D., & Berman, A.J.
Behavioral interaction in social-deprivation reared M. mulatta:
Effects of cerebellar lesions on aggressive and affiliative
behaviors. Journal of Medical Primatology, 1979, 8, 18-28.

    Maintaining isolation-reared adult female Macaca mulatta
    in a group enclosure resulted in rapid resocialization.
    The influence of cerebellar lesions on social behaviors
    was most marked on aggressive interactions and cage
    stereotyped pacing and circling.

313. Lancaster, J.B. Play mothering: The relations between
juvenile females and young infants among free-ranging vervet
monkeys (Cercopithecus aethiops). Folia Primatologia, 1971, 15,
161-182.

    In a study of social behavior of free-ranging vervet
    monkeys living along the Zambezi River near Livingstone,
    Zambia, 295 observations were made in which a juvenile
    female directed some type of maternal behavior toward
    an infant. Juvenile females showed a high degree of
    interest in young infants and would touch, cuddle,
    carry and groom infants whenever they could. This
    opportunity to care for infants provides juvenile females
    with situations in which they can practice not only
    motor skills that are important in maternal behavior but
    also playing the maternal role itself.

314. Lancaster, J.B. Sex and gender in evolutionary perspective.
In   H.A. Katchadourian (Ed.), Human Sexuality: A comparative
and developmental perspective. Berkeley: University of California
Press, 1979, 51-82.

    This chapter summarizes recent data drawn from long-
    term field studies of primates. These new data were
    developed by using careful techniques of sampling
    the behavior of all individuals in a social system,
    and have important implications for the understanding
    of the role of sexual behavior in intergrating primate
    societies, the importance of orgasm in female primate

sexual behavior, the roles played by dominance and
personal preference in mate selection in primate
societies, and the question of avoidance of incestuous
matings. The implications for the evolution of human
sexual behavior are discussed.

315. Lashley, K.S., & Watson, J.B. Notes on the development of
a young monkey. Journal of Animal Behavior, 1913, 3, 114-119.

This early article describes behavior observed in the
development of a young monkey.

316. Lemmon, W.B. Deprivation and enrichment in the development
of primates. Proceedings of the 3rd International Congress of
Primatology, Zurich 1970, 3, 108-115 (Karger/Basel, 1971).

This article presents a comparison of the development of
behavior in chimpanzee infants reared under a variety
of conditions. Some infants were left with their mothers
through weaning, others separated on the day of birth,
and some removed at varying ages within their first
year. All separated infants were placed in a human
home and reared in species isolation. Several of the
separated, human-reared infants were then reintroduced
into the chimpanzee colony and their behavior was compared
with that of infants raised in the colony. This report
focuses only on the most obvious differences between
the three groups.

317. Lemmon, W.B. Experimental factors and sexual behavior in
male chimpanzees. In E.I. Goldsmith & J. Moor-Jankowski (Eds.),
Medical Primatology, 1970, (Karger/Basel) 1971, 432-440.

Male sexuality in the chimpanzee develops effectively
when at least three developmental contingencies are met.
The male should develop a relationship with a mother or
surrogate in infancy sufficiently effective that inca-
pacitating stereotypies can be avoided. The infant and
juvenile should have an adequate opportunity to experience
socialization, including dominance interactions, with
other chimpanzees or human surrogates. Probably no
infantile or juvenile sexual experience is required if
the sexual initiation at puberty is with an adolescent

female who may herself be nulliparous but experienced.
The most effective way to develop adequate socialization
is through play. Play is the obvious situation in
which the innate motor and perceptual behaviors become
integrated and socialized into species-survival
effectiveness.

318. Lemmon, W.B., Temerlin, J. & Savage, E.S. The development
of human-oriented courtship behavior in a human-reared chimpanzee
(Pan troglodytes). In S. Kondo, M. Kawai, & A. Ehara (Eds.),
Contemporary primatology: Proceedings of the 5th International
Congress of Primatology, Nagoya, 1975, 292-294, (Karger/Basel)

A female chimpanzee has been raised to sexual maturity
without contact with her own species. During periods
of maximal gluteal inflation she directs intense quasi-
sexual greeting behaviors toward humans who are not
members of her family. These behaviors include an open-
mouth-cover with panting and pelvic thrusting.

319. Levison, C.A. The development of head banging in a young
rhesus monkey. American Journal of Mental Deficiency, 1970, 75(3),
323-328.

This report describes the development of head banging
in a rhesus monkey which had been reared under conditions
of early social and visual deprivation, but which had,
after release from deprivation, exhibited no stereotyped
behaviors. Head banging developed as a consequence
of a particular set of interactions with the experimenter
and subsequently generalized to other related areas
of behavior.

320. Levison, C.A., & Levison, P.K. Effects of early visual
conditions on stimulation-seeking behavior in young rhesus monkeys,
II. Psychonomic Science, 1971, 22(3), 145-147.

Responding for complex visual stimuli in rhesus monkeys
reared under different early visual conditions showed
persisting effects of early rearing conditions at
later ages. Animals reared for a period after birth
without pattern vision showed less stimulation seeking

Than did a monkey reared with unrestricted access to
perceptual stimuli. Limited enrichment provided by a
10 minute daily exposure to a laboratory environment
of an otherwise visual pattern-restricted animal also
appeared to elevate the monkey's stimulation-seeking
behavior. Type of stimuli affected rate of responding
in all but the younger animals in the restricted group.

321. Levison, C.A., Levison, P.K., & Norton, H.P. Effects of
early visual conditions on stimulation-seeking behavior in infant
rhesus monkeys. Psychonomic Science, 1968, 11(3), 101-102.

Operant responding for complex visual stimuli showed
expected changes as a function of visual deprivation
manipulations in infant rhesus monkeys reared with
normal pattern vision. Infants reared with restricted
access to patterned light showed no systematic changes
consequent to the same manipulations.

322. Levison, C.A., Levison, P.K., & Norton, H.P. Effects of
early visual pattern deprivation on later motor development in
chair-reared rhesus monkeys. Proceedings of the 78th Annual APA
Convention, 1970, 5, 193-194.

Developmentalists have long been interested in the
effects of early physical restriction on the acquisition
of various motor skills (e.g., Dennis, 1940). A rearing
experiment with infant rhesus monkeys suggests that
relative amount of early visual experience may be a
factor in the development of motor skills when physical
restrictions are also present. Patterns of physical
activity obviously involve sensory as well as motor
components, including visual factors in directing the
sighted organism in its physical environment. Early
experience with a "normal visual environment" appears
to have a "priming" function, which is asserted in
later motor performance in infant rhesus monkeys which
are physically restricted from birth.

323. Lewis, J.K., & McKinney, W.T. The effect of electrically
induced convulsion on the behavior of normal and abnormal rhesus
monkeys. Diseases of the Nervious System, 1976, 37(12), 687-693.

The social behavior of sex juvenile rhesus monkeys was
studied in a playroom before, during, and after a series
of electrically induced convulsive (EIC) treatments.
Three subjects had been reared in a normal environment,
and showed the usual levels of social activity for monkeys
of this age. Three were socially deprived early in life
and showed higher levels of self-directed behaviors and
lower levels of social behaviors. At different times
all animals received both EIC and sham EIC three times
per week for 4 weeks, and were observed in a playroom.
In general the experimental subjects showed an increase
in environmental activity as a consequence of EIC
treatments while the control subjects showed decreases
in environmental activity and in several social behaviors.
The control subjects showed clear increases in self-
disturbance behaviors while the experimental animals
tended to show a decrease. The patterns of these changes
thus showed clearly different characteristics of response
to EIC and sham EIC as a function of early rearing
condition. The results are discussed in terms of
possible models for further study of the effects of EIC
on physiological and social variables in a controlled
laboratory setting.

324. Lewis, J.K., McKinney, W.T. Jr., Young, L.D., & Kraemer, G.W.
Mother-infant separation in rhesus monkeys as a model of human
depression: A reconsideration. Archives of General Psychiatry,
1976, 33, 699-705.

Nineteen rhesus monkeys between the ages of 5.9 and 8.5
months were separated from their mothers in five different
studies. While in two of the studies, data indicated
behavioral responses roughly parallel to Bowlby's
protest-despair response to maternal separations, data
across all five studies were sufficiently variable to
bring this technique into serious question as a reliable
and predictable animal model for neurobiologic and
rehabilitation studies.

325. Lichstein, L., & Sackett, G.P. Reactions by differentially
raised rhesus monkeys to noxious stimulation. Developmental
Pshchobiology, 1971, 4(4), 339-352.

Adult monkeys reared in social-sensory isolation were
compared with socially raised controls in 2 studies

assessing aversive reactions to electric shock. In
Experiment 1, the procedure involved competition between
self-paced thirst and shock avoidance, presenting the
monkey with a situation in which it had to absorb 2-sec
of mouth-shock on a drinking tube to obtain water.
When the monkey went through a response cycle by over-
coming a given shock level, taking a variable number of
unshocked drinks, and going 3 min. without touching the
drinking tube, the current was increased one step in
a range from 0.1-2.5 mA. On measures of fluid intake
and time per response cycle isolate monkeys did not differ
from controls, suggesting that thirst motivation did
not differ between groups. On shock trials isolates
tolerated higher levels of shock than controls, even
though they were initially more reactive to very low
current levels. Isolates also showed more negative
generalized effects of shock on nonshocked drinking
tube contacts at a time when controls showed almost no
generalized effects. In Experiment 2, neither isolates
nor controls preferred to drink from a shocked over a
nonshocked tube in a 2-choice free drinking situation,
indicating that shock was a noxious stimulus for both
groups.

326.   Linksley, D.F., Wendt, R.H., Fugett, R., Lindsley, D.B., &
Adey, W.R. Diurnal activity cycles in monkeys under prolonged
visual-pattern deprivation. Journal of Comparative and
Physiological Psychology, 1962, 55(4), 633-640.

Two macaque monkeys, one rhesus and one cynomolgus, were
raised, from 3 weeks of age for 1 year, in an environment
isolated from patterned auditory stimulation by white
noise and deprived of light except for 1 hour of
unpatterned light stimulation each day. The activity
of the two monkeys was continuously monitored by a
stabilimeter system. Analysis of the records revealed
intrinsic diurnal activity cycles which were anchored
to the period of light stimulation and shifted position
on an absolute time scale when the light-stimulation
period was shifted. Dropping of feeding periods or
shifting the feeding schedule had no such effect on the
activity cycle. With each shift in the light schedule
a transition period of 3 to 5 weeks was required before
the activity cycle stablized in a new location relative
to the light.

327. Lindsley, D.B., Wendt, R.H., Lindsley, D.F., Fox, S.S., Howell,
J., & Adey, W.R. Diurnal activity, behavior, and EEG responses in
visually deprived monkeys. Annals of the New York Academy of
Science, 1964, 117, 564-588.

    The work described here was undertaken for purposes other
    than the study of circadian rhythms, but as the results
    show it bears significantly upon the problem of diurnal
    activity and the manner in which it may be influenced
    by light stimulation. Additionally, it deals with
    behavioral and neural changes related to long-term
    light deprivation. Four kinds of experimental observations
    were made: 1) diurnal activity cycles in light-deprived
    monkeys; 2) bar-pressing (motivation) for light stimulation
    in light deprived monkeys; 3) qualitative behavioral
    changes in long-term, light-deprived monkeys; and
    4) brain electrical activity in long-term, light-deprived
    monkeys.

328. Locke, K.D., Locke, E.A., Morgan, G.A., & Zimmerman, R.R.
Dimensions of social interaction among infant rhesus monkeys.
Psychological Reports, 1964, 15, 339-349.

    To determine the basic dimensions of social interactions
    in infant rhesus monkeys, 12 animals were divided into
    groups of four each and observed in a playroom situation.
    Each monkey was observed for a total of 3 hours in 18
    test sessions over a 2-month period. A factor analysis
    of three separate sets of data yielded two independent
    factors: an approach-avoidance factor in which the
    monkey being observed passes or approaches another animal
    who withdraws, and an avoidance-approach factor in which
    the monkey being observed withdraws as another monkey
    passes or approaches. The first factor appears similar
    to what is ordinarily called "dominance" among humans
    and the second factor appears similar to what is ordinarily
    called "submission". Unlike studies with humans, no
    "love-hostility" factor emerged. This was thought to
    be due to the age and/or early social deprivation of Ss.
    There was little consistency in the behavior of the monkeys
    when group compositions were changed.

329. Luttge, W.G. The role of gonadal hormones in the sexual
behavior of the rhesus monkey and human: A literature survey.
Archives of Sexual Behavior, 1971, 1(1), 61-68.

   The importance of the gonads for the display of sexual
   behavior has been realized for hundreds of years, yet
   only within the last century has active research in
   this area been conducted. The present communication
   surveys the literature concerned with the role of
   gonadal hormones in the activation, maintenance, and
   control of sexual behavior in the rhesus monkey (Macaca
   mulatta) and human. To place this topic in perspective
   the survey begins with a description of the normative
   aspects of sexual behavior and the endocrine events
   which correlate with the cyclic behaviors of the female.
   Building on this base the survey then focuses on
   descriptions of the effects on sexual behavior of the
   removal of hormones in adulthood and of the replacement
   of these hormones with either hetero- or homotypic
   hormones.

330. Maple, T. Unusual sexual behaviors of nonhuman primates.
In   J. Money and H. Musaph (Eds.), Handbook of sexology.
Amsterdam: Elsevier, 1977, 1167-1186.

   Unusual sexual behavior in animals occurs infrequently,
   is not well described, well known or well understood.
   The incidence of behavior such as homosexuality, lack
   of sexual interest, self-sucking, masturbation, object-
   directed sexuality, birth-induced sexual responses, and
   the generalization of sexual behavior to alien species
   varies according to the individual, region, and species.
   Most of these types of behavior have been observed in
   both captive and free-ranging animals, and can be said
   to be functional in certain respects. Since all primates
   are socially flexible, the universality of these
   behaviors may be generalized to Homo sapiens. The
   study of unusual sexual patterns in nonhuman primates
   may provide relevant models for the better understanding
   of human sexual behavior. Much has been learned from
   studies of nonhuman primates, but more research is
   required if we are to completely understand the
   relationship between our primate heritage and our
   human nature.

331. Maple, T., & Westlund, B. Interspecies dyadic attachment
before and after group experience. Primates, 1977, 18(2), 379-386.

     Social behavior between representative members of different
     taxa has been observed in the field and studied in the
     laboratory. In the study described here, two female
     rhesus monkeys were paired with female baboons during
     infancy. Preference tests were conducted after inter-
     species pair-living, followed by one month of group
     experience and a subsequent preference test. The
     results indicated that each animal formed social attach-
     ments to the alien partner. Further studies of inter-
     species interactions should contribute to our under-
     standing of behavioral isolation and the process of
     speciation.

332. Maple, T., Brandt, E.M., & Mitchell, G. Separation of
preadolescents from infant rhesus monkeys. Primates, 1975, 16(2),
141-153.

     Each of eight infant rhesus monkeys was paired with a
     preadolescent conspecific for two months and then
     separated. Four of the infants were mother-reared and
     four were isolate-reared. Separation responses were
     compared with data from preseparation and reunion
     phases of the study for all pairs. The results indicate
     that 1) although males interact with infants in a parental
     fashion, preadolescent females show a greater capacity
     for parental behavior, 2) both preadolescents and normal
     (mother-reared) infants contribute to the development of
     a social bond and were relatively less-valued as social
     partners by their preadolescent cage-mates.

333. Maple, T., Erwin, J., & Mitchell, G. Separation of adult
heterosexual pairs of rhesus monkeys: The effect of female cycle
phase. Journal of Behavioral Science, 1974, 2(2), 81-86.

     Four heterosexual pairs of adult rhesus monkeys (Macaca
     mulatta) were divided into two groups: those united at
     the onset of menstrual bleeding by the female member
     of the pair, and those united on the day when ovulation
     was expected. The animals were observed during a
     seven-day period consisting of three distinct phases:
     Preseparation (union), separation and post-separation
     (reunion). Differences were observed according to the

conditions of the females, with males showing different
forms of affiliative behaviors depending on the
condition of the female and females showing greater
interest in the ovulation condition.

334. Maple, T., Erwin, J., & Mitchell, G. Sexually aroused self-
aggression in a socialized adult male monkey. <u>Archives of Sexual</u>
<u>Behavior</u>, 1974, <u>3</u>(5), 471-475.

An abnormal but apparently successful sexual posture
was observed in an adult male rhesus monkey in which
the subject bit its hands and leg in the process of
dismounting. This behavior followed each of several
mounts but ceased toward the end of the series,
culminating in ejaculation. The animal's rearing
history is included, suggesting that early partial
isolation may produce long-lasting abnormalities which,
while unusual, need not seriously affect successful
copulation.

335. Maple, T. , Risse, G., & Mitchell, G. Separation of adult
males from adult female rhesus monkeys (<u>Macaca mulatta</u>) after a
short-term attachment. <u>Journal of Behavioral Science</u>, 1973, <u>1</u>(5),
327-336.

This paper is concerned with the nature of the emotional
bond between adult male and female rhesus monkeys.
Four adult heterosexual pairs were studied. Each pair
lived together for two days, was separated for two
days, and was then reunited for two days. Males appeared
to be more disturbed by the separations than the females
but only the females emitted distress calls (coo vocalizations). The adult heterosexual separations are discussed
in relation to mother-infant separations and to juvenile-
juvenile separations.

336. Marler, P., & Gordon, A. The social environment of infant
macaques. In   P. Glass (Ed.), <u>Environmental influences</u>. New
York: Rockefeller University Press, 1968, 582-589.

This chapter discusses, from a zoological perspective,
Mason's arousal hypothesis as an explanation for behavioral
interactions between infant nonhuman primates and their

mothers. Briefly, Mason's theory states that actions
such as clinging, sucking, and grooming are regarded
as arousal-reducing, while another set, which becomes
predominant later in infant life, serves to increase
arousal. The authors question whether this hypothesis
is broad enough to explain all observed behavioral
patterning, and postulate the existence of endogenous
constraints on the behavior of the developing animal.
Two lines of evidence are reviewed which reflect the
durability of the mother-infant bond: one concerned
with dominance behavior, and the other with the development
of feeding habits. These two areas are discussed with
respect to initial predispositions or appetites which
may affect the behavior of nonhuman primate infants.

337. Mason, W.A. The effects of social restriction on the behavior
of rhesus monkeys: I. Free social behavior. Journal of Comparative
and Physiological Psychology, 1960, 53, 582-589.

    Comparisons were made of the spontaneous social inter-
    actions of monkeys raised in a socially restricted
    laboratory environment and Feral monkeys captured in
    the field. Pairs of Restricted monkeys showed more
    frequent and prolonged fighting and fewer and less
    prolonged grooming episodes than Feral pairs. Differences
    between groups were found in the frequency, duration
    and integration of sexual behavior, which were particularly
    evident in the behavior of males. Restricted and Feral
    males were subsequently tested with the same socially
    experienced females, thus eliminating inadequacies
    in the sexual partner as a factor contributing to the
    differences in male sexual performance. Gross differences
    in the organization of the male copulatory pattern were
    still apparent. In addition to differences between
    groups in the form and frequency of these basic social
    responses, the data suggest that responses to social
    cues are poorly established in monkeys with restricted
    socialization experience.

338. Mason, W.A. Socially mediated reduction in emotional responses
of young rhesus monkeys. Journal of Abnormal and Social Psychology,
1960, 60(1), 100-104.

    Two studies investigated the effect of various social
    stimuli upon emotional responsiveness of 12 young rhesus

monkeys raised in the laboratory in the visual presence
of other young monkeys. Experiment 1 compared the
effect of three stimulus conditions: A familiar partner,
an unfamilar age-peer, and the empty detention cage,
and Experiment 2 extended the range of test stimuli
to include an adult monkey and an albino rabbit. In
both experiments responses indicative of emotional
disturbance were less frequent in the presence of the
partner and the unfamiliar age-peer. These results
suggest that the nature of Ss' previous social experience
was an important determinant of stimulus effectiveness
in reducing emotional distress. The capacity of a
social stimulus to mitigate emotional disturbance
did not appear to be dependent upon nor derived from
feeding or other nurturant experiences. Evidence
was presented suggesting that the adult monkey was a
source of emotional disturbance, producing an effect
similar to that observed in young children and chimpanzees
in the presence of strangers.

339. Mason, W.A. Effects of age and stimulus characteristics on
manipulatory responsiveness of monkeys raised in a restricted
environment. Journal of Genetic Psychology, 1961, 99, 301-308.

The development of manipulatory behavior and its relation
to stimulus characteristics was investigated in two
rhesus monkeys, housed and tested in enclosed isolation
cages during the period from birth to 90 days of age.
The stimulus objects, each representing one of three
classes, movable, flashing, and stationary, were presented
singly in successive 5-day periods. Manipulatory responses
appeared in both Ss by the second day of life. Response
frequencies were generally low, however, during the
first 15 days and subsequently increased rapidly and
progressively for both Ss. Responses were highest with
movable stimuli, next highest with flashing stimuli
and lowest with stationary stimuli. The data indicate
that responses are generally more frequent on the first
day of a 5-day exposure period.

340. Mason, W.A. The effects of social restriction on the
behavior of rhesus monkeys: II. Tests of gregariousness. Journal
of Comparative and Physiological Psychology, 1961, 54(3), 287-290.

The gregarious behavior of monkeys raised in the laboratory
(Restricted group) and monkeys captured in the field
(Feral group) was investigated in several situations
by presenting a choice between social and non-social
alternatives. Pairs of Restricted monkeys made less
social choices and fought more frequently following a
social choice than did pairs of Feral monkeys. Restricted
and Feral males were frequently tested with the same
socially experienced female incentive animals, and the
number of social choices by Restricted males increased
sharply as compared with their performance in the first
experiment. When given an opportunity to choose between
Feral males and Restricted males, socially experienced
females uniformly preferred the Feral subjects. The
results support the conclusion that orderly and harmonious
intraspecies social relations in rhesus monkeys are
dependent on previous socialization experience. Socially
restricted monkeys are apparently not preferred either
by individuals of similar social history or by monkeys
born in the field.

341. Mason, W.A. The effects of social restriction on the behavior
of rhesus monkeys: III. Dominance tests. Journal of Comparative
and Physiological Psychology, 1961, 54(6), 694-699.

Dominance relationships within a group of six monkeys
reared in the laboratory (Restricted monkeys) were compared
with dominance relations within a group of six monkeys
captured in the field (Feral monkeys). Stable dominance
relationships were evident between pairs of Feral monkeys
before the introduction of food-competition tests, and
there was little change in dominance status between
the beginning and the end of competitive testing. During
competitive interaction, fighting was infrequent and
was initiated only by dominant monkeys. Noncompetitive
social interactions of Restricted monkeys provided
little indication of subsequent status in food-competition tests. During competitive interaction dominance
relations were fluctuating and unstable and dominance
was unequivocally established in only 7 of 15 Restricted
pairs. Fighting was frequent during competitive interaction
and was displayed by subordinate as well as by dominant
monkeys. These data emphasize the importance of social
learning in the establishment and maintenance of
dominance relations.

342. Mason, W.A. Social development of rhesus monkeys with
restricted social experience. Perceptual and Motor Skills, 1963,
16, 263-270.

   Social development was studied in six pairs of rhesus
   monkeys whose only opportunity for physical contact
   with other monkeys was in brief test sessions. The
   animals were observed during the period from 25 to 85
   days of age (Phase I) and observations were resumed 120
   days later for 30 additional days (Phase II). The
   dominant social response initially was clasping;
   occasionally this was accompanied by mouthing and by
   thrusting. The incidence of clasping declined steadily
   and there was a corresponding growth in the frequency
   of social play. The mean age of onset for play was
   40 days. At the beginning of the second phase of testing,
   clasping was again the dominant activity; but it was
   quickly surpassed by play. Aggression and adult forms
   of grooming and sexual behavior were never observed.

343. Mason, W.A. Determinants of social behavior in young
chimpanzees. In H.F. Harlow & F Stolnitz (Eds.), Behavior of
nonhuman primates (Vol. 2). New York: Academic Press, 1965,
335-364.

   The development of the infant chimpanzee is discussed
   with regard to social responsiveness under various levels
   of stimulation. The concept of arousal is investigated
   by examining the nonspecificity of stimulation, qualitative
   changes in behavior as a function of quantitative stimulus
   change, and the reinforcement value of stimulation
   to the social development of the infant primate.

344. Mason, W.A. The social development of monkeys and apes. In
I. Devore (Ed.), Primate behavior. New York: Holt, Rinehart, and
Winston, 1965, 514-543.

   This article presents a general overview of primate
   ontogeny under feral and laboratory conditions. The
   normative development of the infant-mother, maternal,
   peer, and heterosexual affectional systems are discussed
   as well as the effects of social deprivation on these
   behavior patterns.

345. Mason, W.A. The effects of environmental restriction on the
social development of rhesus monkeys. In C.W. Southwick (Ed.),
Primate social behavior. New York: Van Nostrand Reinhold Co.,
1967, 161-173.

   This paper describes a series of experiments designed to
   provide a broad analysis of the effects of artificially
   restricting opportunities for social learning during
   the first few years of life. The major comparisons
   involved two groups of rhesus monkeys (Macaca mulatta),
   each about 2½ years of age. One group of three males
   and three females was separated from their mothers
   shortly after birth and were raised in individual
   wire cages which permitted visual and auditory stimulation,
   but no physical contact. The second group of three
   males and three females had been captured in the wild
   and brought to the laboratory at 20 months of age. One
   month before the beginning of the experiment these six
   animals were housed in the same manner as the first
   group. The results of this investigation are discussed
   with regard to various social behaviors and how they are
   affected by social deprivation.

346. Mason, W.A. Motivational aspects of social responsiveness
in young chimpanzees. In W. Stevenson, E.H. Hess, & H.L. Reingold
(Eds.), Early behavior. New York: Wiley, 1967, 103-126.

   The importance of clinging, grooming, and play are
   discussed with regard to social interaction and socialization
   of the developing chimpanzee. The motivational parameters
   of these behaviors are examined via experimentation with
   ambient noise levels, novel surroundings, strange companions, physical restraint, separation from cagemates,
   and drugs.

347. Mason, W.A. Early social deprivation in the nonhuman primates:
Implications for human behavior. In D.C. Glass (Ed.), Environ-
mental influence. New York: The Rockefeller University Press &
Russell Sage Foundation, 1968, 70-112.

   This article presents a general examination of the
   nonhuman primate as a model for the development of
   various behavior patterns during infancy and childhood
   in man. The development of filial responses as a means
   of reducing stimulation early in life and the onset of

exploitative arousal increasing behaviors later in life
are discussed in light of normal environmental conditions
and deprivation states.

348. Mason, W.A. Scope and potential of primate research. In
J.H. Masserman (Ed.), Science and psychoanalysis (Vol. 12) Animal
and human. New York: Grune and Stratton, 1968, 101-118.

The use of nonhuman primates to study both normal and
psychopathological development under various environmental
conditions is discussed with regard to human ontogeny.
The nonhuman primate is analyzed as a model for man, a
conceptual model of man, and a comparative-evolutionary
model for the knowledge of environmental adaptation.

349. Mason, W.A. Early deprivation in biological perspective.
In V.H. Denenberg (Ed.), Education of the infant and young child,
New York: Academic Press, 1970, 25-50.

In this chapter a biological perspective on early deprivation
is applied to problems of human education. Differences
between the information processing systems involved
in so-called "instinctive" and "intelligent" behaviors
are discussed, and the author attempts to show that what
most distinguishes the intelligent from the instinctive
organism is the degree of openness of their information
processing systems. This openness is not seen as creating
intelligence, rather it establishes the conditions in
which the growth of intelligence is possible. A biological perspective can contribute to both the recognition
of the accumulative nature of intellectual growth, and
to the appreciation of how the higher levels of intellectual functioning are critically dependent upon the prior
development and the integrity of the levels that lie
beneath them.

350. Mason, W.A. Information processing and experiential
deprivation: A biological perspective. In   F.A. Young and
D.B. Lindsley (Eds.), Early experience on visual information
processing in perceptual and reading disorders. Washington,
D.C.: National Academy of Sciences, 1970, 302-323.

Our appreciation of the special skills of the human organism
as an information-processing system and our understanding
of the disabilities to which this system is liable will
be improved if they are considered within a broad biologic
perspective. This chapter discusses the general trend
in mammalian evolution for schemata (representational
processes) to become more open, more responsive to
variations in environmental input. In the more advanced
species, early environmental restriction will affect
schemata in two ways: It will impede their complete
development and elaboration by with-holding "essential"
classes of information to which the individual (as a
representative of a particular species) is selectively
"tuned", and it will lead to the development of seemingly
aberrant or distorted schemata that are actually reflections
of (or isomorphic with) the peculiar properties of the
rearing environment.

351. Mason, W.A. Motivational factors in psychosocial development.
Nebraska Symposium on Motivation, 1970, 18, 35-67.

Psychosocial development is viewed with reference to two
major developmental trends, each corresponding with a
fundamental adaptive problem for the growing primate.
The first trend is inferred from the filial contact-
seeking behaviors of the infant towards its mother.
These behaviors are most frequently seen under high
arousal conditions and their performance leads to arousal
reduction. Arousal reduction provides the primary mechanism
of reinforcement in the development of the infant's
tie to the mother. The second major developmental trend
is inferred from exploitative behaviors such as social
play, motor play, and investigatory activities.
Exploitative behaviors are elicited by moderate increments
of arousal and their performance leads to further
increments in arousal. Objects that elicit or sustain
exploitative behaviors are effective reinforcers and
they are primary agents in moving the infant away from
the mother and into a larger world. The syndrome of
abberations and deficiencies seen in primates raised
in social isolation is viewed as manifestations of
these two major developmental trends, operating on
reduced or altered environmental input.

352.   Mason, W.A.   Regulatory functions of arousal in primate
psychosocial development.   In C.R. Carpenter (Ed.), <u>Behavioral
regulators of behavior in primates</u>.   Lewisburg, Pa.:   Bucknell
University Press, 1973.

   This section attempts to account for primate
   psychosocial development within a framework that
   acknowledges the broad similarities in development among
   different species and yet is able to deal with the
   specific effects of experience on the course of
   individual development.   Psychosocial development is
   viewed with reference to two major developmental trends,
   each corresponding to a fundamental adaptive problem
   for the growing primate.   The first trend is inferred
   from the filial contact-seeking behaviors, and provides
   the basis for the adjustment to the mother.   The
   second major developmental trend is inferred from
   exploitive behaviors such as social play, motor play,
   and investigatory activities, and insofar as this trend
   brings the growing primate into contact with other
   individuals, it extends the opportunities and the
   necessity for social learning.   The syndrome of
   aberrations and deficiencies seen in primates raised
   in social isolation is viewed as a manifestation of
   these two major developmental trends operating on
   reduced or altered environmental input.

353.   Mason, W.A.   Social experience and primate cognitive
development.   In G.M. Burghardt and M. Bekoff (Eds.), <u>The
development of behavior: Comparative and evolutionary aspects</u>.
New York: Garland Press, 1978, 233-251.

   This chapter examines the distinction between two
   types of knowing.   The author believes that if we are
   to reach an understanding of the role of social
   experience in the cognitive development of certain
   species, it is essential to distinguish between the
   organism's competences, which are reflected in its
   specific achievements, and its generalized expectancies
   or coping strategies, which are reflected in its basic
   stance toward the environment.   Although both kinds of
   knowledge are dependent on experience, they are seen
   to have different developmental antecedents and carry
   quite different implications for behavioral adaptability.

354. Mason, W.A., & Berkson, G. Conditions influencing vocal
responsiveness of infant chimpanzees. Science, 1962, 137, 127-128.

   Infant chimpanzees were tested to determine the
   effects on distress vocalizations (whimpering,
   screaming) of stimulus conditions approximating the
   hysical relationship to the mother. Under such
   conditions spontaneous vocalizations were infrequent,
   and vocal responsiveness to a painful stimulus was
   substantially reduced.

355. Mason, W.A., & Berkson, G. Effects of maternal mobility on
the development of rocking and other behaviors in rhesus monkeys:
A study with artificial mothers. Developmental Psychobiology,
1975, 8(3) 197-211.

   Mechanically driven mobile artificial mothers effectively
   prevented the development of sterotyped body-rocking
   in rhesus monkeys. Monkeys were maternally separated at
   birth and assigned to 2 groups. Both groups were placed
   with surrogates, identical in construction except that
   for 1 group the surrogate was in motion 50% of the
   time from 0500 hours to 2400 hours each day, and for
   the other group the surrogate was stationary. All but
   1 of the 10 monkeys raised with stationary artificial
   mothers developed rocking as an habitual pattern whereas
   none of the 9 monkeys raised with mobile mothers did so.
   The data also suggest that emotional responsiveness was
   reduced in monkeys raised with mobile mothers, compared
   to monkeys raised with stationary devices.

356. Mason, W.A., & Fitz-Gerald, F.L. Intellectual performance
of an isolation-reared rhesus monkey. Perceptual and Motor Skills,
1962, 15, 594.

   This report describes the performance of a rhesus monkey
   raised in social isolation on four tasks generally
   believed to demonstrate intellectual competence:
   (1) simple object discrimination, (2) discrimination
   reversal, (3) discrimination learning set, and (4) form
   discrimination and generalization. S was maintained from
   birth in an enclosed cage. It was tested for manipulatory
   responsiveness during the first 90 days of life (Mason,
   1961). Solid foods were introduced on Day 121 and

were being readily accepted by Day 141. Testing
was conducted with a modified Wisconsin General Test
Apparatus, attached to the living cage. Training on
object discrimination began when S was 150 days old.
The two objects differed in form, color, and size. S
was tested for 15 days at a rate of 25 trials a day.
At no point during training was daily performance below
92% correct. On Day 16 the problem was reversed and
performance fell to 28% correct. Reversal learning was
rapid, and errorless performance was achieved by the
fifth training day.

357. Mason, W.A., & Green, P.C. The effects of social restriction
on the behavior on rhesus monkeys: IV. Responses to a novel
environment and to an alien species. Journal of Comparative and
Physiological Psychology, 1962, 55(3), 363-368.

Comparisons were made of the reactions of monkeys
reared in the laboratory (Restricted monkeys) and
monkeys captured in the field (Feral monkeys) to an
albino rat and to isolation in an unfamiliar room.
Restricted monkeys, as compared with Feral monkeys,
were not as gentle with their rats, made fewer contacts
with the rat in the living cage, and in subsequent tests
in which the rat was presented as a social incentive,
made fewer social choices. The reactions of Feral and
Restricted monkeys to isolation in an unfamiliar room
contrasted sharply. Restricted monkeys crouched, sucked
thumbs or toes, clasped themselves, and engaged in rocking
or other stereotyped repetitive behaviors. None of these
responses was observed in the Feral group. Feral
animals had higher locomotor scores and more frequently
engaged in gross motor activities (jumping, turning
backward somersaults). Evidence was presented suggesting
that the syndrome of self-directed responses observed
in socially restricted monkeys is derived from infantile
responses ordinarily made with reference to the mother.
In accordance with this interpretation, it would be
expected that a rat, which is appropriate for relatively
simple primate social activities such as holding or
clasping, would be more attractive to Feral monkeys than
to Restricted monkeys for which such responses are
habitually self-directed. This expectation was supported
by the present results.

358. Mason, W.A. & Hollis, J.H. Communication between young
rhesus monkeys. Animal Behaviour, 1962, 10, 211-221.

   The communication performance of 12 rhesus monkeys
   was investigated in a situation in which rewards of both
   members of a pair of monkeys could not exceed chance
   levels unless the operator monkey responded to cues
   provided by the informant monkey which indicated the
   location of food. All animals were born in the laboratory,
   separated from their mothers within 18 hours after birth
   and subsequently housed in individual wire cages.
   Approximately 11 months before the start of the experimentation 6 of these animals were housed in pairs. The
   results of the experiments are discussed with regard to
   social experience and the ability to discriminate social
   cues.

359. Mason, W.A., Kenny, D. Redirection of filial attachments in
rhesus monkeys: Dogs as mother surrogates. Science, 1974, 183,
1209-1211.

   Rhesus infants raised from birth with their mothers, age-
   mates, or cloth surrogates for periods varying from
   1 to 10 months were separated from these objects and
   placed with dogs. Contrary to previous suggestions that
   were consistent with the notions of a critical period
   for attachment formation and irreversibility of filial
   bonds, the monkeys formed strong and specific attachments
   to their canine surrogates.

360. Mason, W.A. & Riopelle, A.J. Comparative psychology. In
P.R. Farnsworth (Ed.) Annual review of psychology, (Vol. 15),
1964, 143-180.

   This article presents a very extensive review of primate
   research encompassing areas of learning, motivation,
   and social behavior under feral deprivation
   conditions.

361. Mason, W.A., & Sponholz, R.R.   Behavior of rhesus monkeys
raised in isolation. Journal of Psychiatric Research, 1963, 1(4),
299-306.

    Two rhesus monkeys (isolates) were raised in enclosed
    isolation cages from birth until early adolescence
    (16 months). Two additional monkeys of the same age,
    raised routinely in wire mesh cages, were used as a
    comparison group (restricted monkeys). Isolates were
    paired with each other and with restricted monkeys in
    various phases of the experiment and in all received
    more than 1000 hours of exposure to the test situations.
    The isolates appeared to be traumatized by the extra-cage
    environment. Crouching was their characteristic posture
    throughout the experiment. Few responses were directed
    toward other animals or the physical environment and
    the most common reactions to social contact were
    submission or flight.

362. Mason, W.A., Blazek, N.C., & Harlow, H.F.   Learning capacities
of the infant rhesus monkey. Journal of Comparative and Physiological Psychology, 1956, 49(5), 449-453.

    The performance of six infant rhesus monkeys was studied
    on tests of discrimination learning, delayed response, and
    patterned strings during the first year of life. Individual
    discrimination learning problems were readily mastered
    at approximately 150 days of age, and interproblem
    learning, or the ability to form learning sets, was
    subsequently demonstrated, although the rate of improve-
    ment was much slower than that shown by adult rhesus
    monkeys. The young monkey is capable of successful
    delayed-response performance with delay intervals up to
    40 sec., but the level of proficiency is probably below
    that attained by adult monkeys. Solution of parallel and
    converging string problems proved to be relatively easy.
    Considerable difficulty was apparent when the crossed
    patterns were introduced, and a high degree of
    negative transfer was indicated.

363. Mason, W.A., Davenport, R.K., Jr., & Menzel, E.W., Jr. Early
experience and the social development of rhesus monkeys and
chimpanzees. In G. Newton & S. Levine (Eds.), Early experiences
and behavior. Springfield, Illinois: C.C. Thomas, 1968, 440-480.

    This paper discusses various primate studies relating
    to early experience and social development. Comparisons
    are made between feral animals and animals raised under
    social deprivation or isolation conditions.

364. Mason, W.A., Green, P.C. & Posepanko, C.J. Sex differences
in affective-social responses of rhesus monkeys. Behavior, 1960,
16(1-2), 74-83.

    These researches investigated the stereotyped
    affective-social responses of adult male and female
    rhesus monkeys to human observers and to a situation
    designed to produce conflict. In the first experiment
    threat responses to the human observer were scored on a
    four point scale by three observers working independently.
    Each observer obtained significantly more frequent and
    intense threat reactions from female than from male
    subjects. In a second experiment, food was presented
    in proximity to an aversive stimulus and records were
    obtained of threat reactions, lipsmacking, fear grimaces,
    and food acceptance. Females displayed significantly
    more threat responses and fear grimaces. Differences
    between sexes in lipsmacking and food acceptance were
    in the expected direction, but are not statistically
    significant.

365. Mason, W.A., Hill, S.D., & Thomsen, C.E. Perceptual factors
in the development of filial attachment. Proceedings of the
3rd International Congress of Primatology, 1971, 3, 125-133.

    Infant rhesus monkeys (Macaca mulatta) were separated
    from their mothers at birth and housed in cages with a
    distinctive red-and-white checkered interior. Each
    cage contained a simple artificial mother consisting
    of a block of wood covered with blue-green acrylic fur.
    Beginning at approximately 2 weeks of age and at monthly
    intervals thereafter each monkey was tested in a variety
    of situations designed to measure different aspects of
    responsiveness to the familiar surrogate and the familiar
    environment as well as different environments and

unfamiliar surrogates. The results of this
investigation are discussed with regard to distress
vocalizations and changes in heart rate.

366.   Mason, W.A., Hill, S.D., & Thompson, C.E. Perceptual aspects
of filial attachment in monkeys. In N.F. White (Ed.), Ethology
and psychiatry. Toronto: University of Toronto Press, 1974, 84-
93.

    Eighteen infant rhesus monkeys (Macaca mulatta) were
    separated from their mothers on the first day of life and
    reared in individual cages with a distinctive red and
    white checkered interior. Each cage contained a simple
    surrogate mother covered with blue-green acrylic fur.
    Animals were exposed to familiar or unfamiliar surroundings
    and surrogates and distress vocalizations to these stimuli
    were recorded.

367.   Mason, W.A., Hollis, J.H. & Sharpe, L.G.   Differential
responses of chimpanzees to social stimulation. Journal of
Comparative and Physiological Psychology, 1962, 55(6), 1105-1110.

    Two experiments investigated the responses of young
    chimpanzees to various forms of social stimulation
    presented by a stimulus-person (S-P) acting in
    accordance with a predetermined role. The social
    activities included play, petting, grooming S, and
    soliciting grooming from S. Social stimuli were
    presented singly in Experiment 1, and the primary
    measure of preference was proximity to the S-P.
    Experiment 2 employed a paired-comparison technique
    and preference was measured by frequency of window
    openings giving access to S-P. The ordering of
    social activities obtained from the two experiments
    were in good agreement. Play was most preferred; petting
    and being groomed occupied intermediate positions;
    and grooming S-P was least preferred. The results clearly
    show that specific forms of social stimulation similar
    to those commonly occurring in chimpanzee social
    interactions have definite and differential rewarding
    effects and may therefore be regarded as relevant factors
    in the development and maintenance of companionship
    preferences, and in the process of social control.

368.   Mason, W.A., Saxon, S.V., & Sharpe, L.   Preferential responses of young chimpanzees to food and social rewards. Psychological Record, 1963, 13, 341-345.

        Three young chimpanzees were permitted to choose between
        social rewards presented by humans and food. In the
        first experiment the independent variables were type of
        social reward (play or petting) and degree of food
        deprivation. Play was chosen consistently more than
        petting, and the effect of food deprivation was to reduce
        the frequency of social choices. Under low deprivation,
        two Ss selected play reliably more often than food. In
        the second experiment the same social rewards were compared
        with food of high, moderate, and low preference. Play
        was again more effective than petting, and the frequency
        of social choices increased with decreasing food
        preference. Two Ss preferred play over food under all
        conditions.

369.   Masserman, J.H., & Pechtel, C.   Neuroses in monkeys: A
preliminary report of experimental observations. Annals of the New
York Academy of Science, 1953, 56, 253-265.

        This report discusses experimental neuroses in monkeys.
        The animals were subjected to a "psychologically tramatic"
        conflict between learned feeding patterns and presumably
        "innate" reactions of fear to the exhibition of a toy
        snake. All of the animals developed experimental
        neuroses characterized by various inhibitions, regressions,
        phobias, compulsions, organic dysfunctions, neuro-
        muscular disabilities, sexual deviations, and alterations
        in social relationships that were sufficiently
        persistent and progressive to interfere seriously
        with the health and well-being of the subjects. The
        therapeutic procedures employed were effective only
        when they relieved the etiologic motivational conflict
        by methods specifically suited to the individual
        characteristics and unique experiences of each animal.

370.   Maxim, P.E., Bowden, D.M. & Sackett, G.P.   Ultradian rhythms
of solitary and social behavior in rhesus monkeys. Physiology
and Behavior, 1976, 17, 337-344.

        Day-long observations were made on the social and
        nonsocial behaviors of five subadult to adult male

monkeys. Social interaction between pairs of monkeys
was found to occur as a cyclic process. In three
of four pairs, the lengths of these cycles occurred
as multiples of 45 min. Four nonsocial behaviors --
--ingestion, self-grooming, exploration and locomotion
were found to occur as cyclic processes when the monkeys
were in either a social or solitary setting. The
lengths of these cycles again occurred as multiples
of 45 min and in all four social pairings exploratory
cycles of individual monkeys appeared to precede and
be closely in phase with social interaction cycles.

371. McCulloch, T.L. The role of clasping activity in adaptive
behavior of the infant chimpanzee: III. The mechanism of reinforcement. Journal of Psychology, 1939, 7, 305-316.

The present report integrates information regarding
approach-clasping in the infant chimpanzee. The
effects of noxious stimuli are discussed with regard
to this behavior as well as the various implications
for reinforcement theory and learning.

372. McCulloch, T.L., & Haslerud, G.M. Affective response of an
infant chimpanzee reared in isolation from its kind. Journal of
Comparative and Physiological Psychology, 1939, 28, 437-445.

An infant chimpanzee, reared in isolation from other
animals, was presented with a variety of stimulus
objects in a food-approach situation. When seven
months of age the subject showed affective disturbance
to moving objects only, most of this disturbance being
identified as avoidance, and a small portion of it as
aggression. When tested again at fifteen months, the
subject showed no differentiation on the basis of
movement; he exhibited more intense disturbance, extended
it to a larger range of objects, and exhibited much more
aggression. In both tests, those objects which elicited
greatest aggression ranked intermediate in respect to
capacity to elicit avoidance responses. The performance
of the subject is compared with that of other, socialized
chimpanzees in a comparable situation.

373. McKinney, W.T. Jr. Animal modeling of psychopathology:
Current status of the field. In J. Westermeyer (Ed.), <u>Anthropology
and mental health</u>. The Hague, Netherlands: Mouton Publishers,
1974, 259-269.

     This paper discusses the criteria involved in
     producing animal models for human psychopathology
     and refers to the major social induction techniques
     which are being used in animals. These include
     (1) social isolation, (2) experimental helplessness,
     and (3) attachment behavior and separation studies.
     Recent approaches to biochemical studies in primate
     models are also outlined.

374. McKinney, W.T. Jr. Animal models in psychiatry. <u>Perspectives
in Biology and Medicine</u>, 1974, <u>17</u>(4), 529-541.

     This paper discusses the relevance of animal models for
     human psychopathology and reviews two areas of primate
     research which provide potential use for this kind of
     model in the field of psychiatry. The first of these
     areas deals with the production of animal models of
     psychoses, the second with studies designed to create
     animal models for affective disorders.

375. McKinney, W.T. Jr. Primate social isolation: Psychiatric
implications. <u>Archives of General Psychiatry</u>, 1974, <u>31</u>, 422-426.

     Social isolation of rhesus monkeys for the first 6
     to 12 months of life produces severe and persistent
     behavioral effects including social withdrawal,
     rocking, huddling, self-clasping, stereotyped
     behaviors, and inappropriate heterosexual and
     maternal behaviors as adults. The mechanisms by
     which these effects are produced are uncertain and
     require additional investigations. The social
     isolation syndrome has been likened to several human
     psychopathological states, but exact labeling of it in
     human terms is premature at present. Rather the
     syndrome should be viewed in terms of its heuristic
     value as a model system for further clarifying the
     interactions among early rearing conditions, their
     possible neurobiological consequences, and subsequent
     social behaviors.

376. McKinney, W.T. Jr. Psychoanalysis revisted in terms of experimental primatology. In E.T. Adelson (Ed.), <u>Sexuality and psychoanalysis</u>, New York: Brunner Mazel, 1975.

This chapter discusses the relationship between psychoanalysis and experimental primatology in three major areas: (1) the effects of differential early rearing conditions on the development of a variety of social behaviors; (2) the operationalization of psychoanalytic concepts; and (3) experimental psychopathology. The major areas of special interest for psychiatrists in primate behavioral research include the development of attachment behaviors, the effects of social isolation, separation studies, the possible experimental simulation of learned helplessness, and the use of primate models to study the interactions between social and biological determinants of behavior.

377. McKinney, W.T. Jr. Animal behavioral/biological models relevant to depressive and affective disorders in humans. In J.G. Schulferbrandt and A. Raskin (Eds.) <u>Depression in childhood: Diagnosis, treatment and conceptual models</u>, New York: Raven Press, 1977.

This chapter describes the rationale for animal models and the basic implications of these models for any form of human psychopathology. Two specific approaches to the area of studying depression in young nonhuman primates are discussed in order to illustrate the potential value of such work.

378. McKinney, W.T. Jr. Biobehavioral models of depression in monkeys. In J. Hanin and E. Usdin (Eds.), <u>Animal models in psychiatry and neurology</u>. New York: Pergamon Press, 1977, 117-126.

This article reviews research on separation in nonhuman primates and outlines some of the animal models for depression used by the author and others at the Wisconsin Primate Laboratory. The animal modeling field is seen as useful for providing systems in which the complex interactions between social and biological variables can be studied and more than just correlations obtained.

379. McKinney, W.T., Jr., & Bunney, W.E. Jr. Animal model of
depression. <u>Archives of General Psychiatry</u>, 1969, <u>21</u>, 240-248.

>The purpose of this paper is to review the data
>relevant to the development of an animal model
>of "depression" and to discuss possible research
>strategies which could be used to create such a
>model. The evidence has been reviewed in this
>paper, which suggests the possibility of creating
>an animal model of depression, comes from two
>main sources: (1) animal separation experiments;
>(2) anecdotal case histories of animals who have
>developed "depressive-like" syndromes. The
>implications for research are discussed under six
>major categories: (1) evaluation of baseline data,
>(2) methods for induction of depression, (3) sensitization
>to depression, (4) methods to evaluate change, (5)
>reversal of depression, and (6) minimal requirements
>for an animal model of depression. The limitations
>of animal experimentation in reference to confirming
>or negating current psychiatric theories are
>recognized. A comparative approach might, however,
>identify some new significant problems for research
>and can provide a system for investigation of
>concepts by allowing them to be tested on an
>observational and experimental level in animals.

380. McKinney, W.T., Jr., Suomi, S.J., & Harlow, H.F. New models
of separation and depression in rhesus monkeys. <u>Proceedings of
the American Association for the Advancement of Science Symposium</u>,
Chicago, Ill., 1970.

>Separation of rhesus monkeys which have formed
>affectional bonds can lead to a severe depressive
>syndrome. In the case of monkeys up to a year of
>age, the reaction occurs in two stages, an initial
>stage of protest and a second stage of despair. Such
>reactions may be useful in creating animal models for
>human depression though it is important to specify the
>various parameters of separation being discussed.
>Variables dealt with in this paper include the object
>from which separated, the situation from which
>removed, the environment during the period of separation,

and the age of the animal. The data are
discussed in relation to the role of separation in
depression.

381. McKinney, W.T., Jr., Suomi, S.J., & Harlow, H.F.   Depression
in primates. American Journal of Psychiatry, 1971, 127(10),
1313-1320.

    The authors present the results of a number of
    experiments designed to produce depressive behavior
    in young rhesus monkeys and outline their plans
    for further experiments with monkeys. These studies
    are part of a research program aimed at creating an
    animal model of depression that should make it
    possible to study the effects of manipulation of the
    social and biological variables that are thought
    to be important in human depression.

382. McKinney, W.T. Jr., Suomi, S.J., & Harlow, H.F.   The sad
ones. Psychology Today, 1971, 4(12), 60-63.

    This article describes research with rhesus monkeys
    which is an attempt to study systematically some of
    the variables thought to be important in depression.
    Specifically, investigators are trying to create
    experimental animal models for human depression since
    such models have the potential of providing an experimental
    background for some of the theoretical frameworks
    within which depression is viewed by clinicians. This
    research was designed to meet the need for a more
    rigorous method of evaluating the widespread
    theories about depression.

383. McKinney, W.T. Jr., Suomi, S.J., & Harlow, H.F.   How they're
using monkeys to study depression. Resident and Staff Physician,
1972, 45-49.

    This article presents a general explanation of the use
    of nonhuman primate models for psychiatric disorders.
    Several techniques (maternal separation, isolation
    in the vertical chamber, and drug induced depression)
    for instituting behavioral anomalies in rhesus monkeys
    are discussed.

384. McKinney, W.T., Jr., Suomi, S.J., & Harlow, H.F. Repetitive
peer separations of juvenile-age rhesus monkeys. Archives
of General Psychiatry, 1972, 27, 200-204.

   Male rhesus monkeys (Macaca mulatta) 3 years of age were
   studied before, during, and after a series of four
   separations from equal-aged peers with whom they had
   formed stable social bonds. The two-week separations
   were associated with increases in locomotion and
   environmental exploration and decreases in passivity.
   There was no suggestion of any "despair" stage as has
   been reported in younger organisms, thereby suggesting
   the importance of age as a variable in determining
   the response to separation.

385. McKinney, W.T., Jr., Suomi, S.J., & Harlow, H.F. Vertical-
chamber confinement of juvenile-age rhesus monkeys. Archives of
General Psychiatry, 1972, 26, 223-228.

   Male rhesus monkeys (Macaca mulatta) 3 years of age
   were studied daily before and after ten weeks of
   confinement in a vertical-chamber apparatus designed
   to facilitate production of psychological disturbance.
   This study represents an initial effort to move beyond
   the use of young monkeys in a depression research
   program. Chamber confinement results in a significant
   increase in contact clingling between animals and a
   decrease in locomotion following removal from the
   apparatus. These behavior patterns are very typical
   of laboratory-reared rhesus monkeys of this age and may
   represent a maturational regression induced by
   social means.

386. McKinney, W.T. Jr., Suomi, S.J., & Harlow, H.F. Methods and
models in primate personality research. In J.C. Westman (Ed.),
Individual differences in children. New York: John Wiley & Sons,
1973, 265-287.

   This chapter reviews the body of knowledge generated
   regarding primate affectional systems. Next, the
   behavioral aberations noted in social isolation and
   mother-infant separation studies are summarized.
   Finally, several techniques devised for manipulating
   personality and producing precise forms of stress are
   described to illustrate the possibility of experimentally

controlling both the past life experience and
the current state of monkey subjects.

387. McKinney, W.T. Jr., Suomi, S.J., & Harlow, H.F.   Experimental
psychopathology in nonhuman primates.   In D.A. Hamburg (Ed.),
American handbook of psychiatry, 2nd Edition, Vol. 6.   New York:
Basic Books, 1975.

This chapter discusses the use of primate models for
understanding human psychopathology. The major areas
of interest for psychiatrists include the development
of kinds of attachment behavior, the effects of
social isolation, separation studies, the possible
experimental simulation of learned helplessness, and
various biological approaches being developed. Animal
data from these areas are presented and discussed within
the framework of a viable animal model for depression
that may facilitate a more comprehensive understanding
of this particular syndrome and enable studies to be
done that are currently impossible to perform utilizing
human beings.

388. McKinney, W.T. Jr., Suomi, S.J., & Harlow, H.F.   New models
of separation and depression in rhesus monkeys. In J. Scott and
E. Senay (Eds.), Separation and depression: Clinical and research
aspects. Washington D.C.: American Association for the Advancement of Science, 1973.

This report discusses an extensive series of
experiments that specifically focused on
separation and its effects in rhesus monkeys.
In an attempt to specify carefully what is meant
by separation, some of the parameters that the
authors have examined and which are discussed
in this paper include: (1) the object from
which an individual is separated; (2) the
situation from which an individual if removed;
(3) the environment during the period of
separation; and (4) the developmental stage or
age at the time of separation.

389. McKinney, W.T. Jr., Kliese, K.A., Suomi, S.J., & Moran, E.C.
Can psychopathology be reinduced in rhesus monkeys? An experimental
investigation of behavioral sensitization. <u>Archives of General
Psychiatry</u>, 1973, <u>29</u>, 630-634.

    The problem investigated was whether one experience of
    profound environmental deprivation early in development
    would influence the response of rhesus monkeys to a
    subsequent period of deprivation. Subjects with a
    history of previous vertical chamber confinement were
    tested 1½ years later and compared with control subjects
    with no prior history of environmental deprivation.
    Monkeys with a history of early chamber confinement
    exhibited residual effects 1½ years after their initial
    experience but before reentering the chamber. Further-
    more, the current chambering experience had a significant
    impact in terms of producing decreased environmental
    activity and increased passivity. Effects of the current
    chambering experience occurred whether or not individual
    subjects had had a prior experience. These data con-
    trast with previous studies which indicate prior
    separation to be a high risk-predisposing event in
    rhesus monkeys.

390. McKinney, W.T., Suomi, S.J., Kliese, K.A., & Moran, E.C.
Can psychopathology be reinduced in rhesus monkeys? <u>Archives of
General Psychiatry</u>, 1973, <u>29</u>, 630-634.

    The problem investigated was whether one experience of
    profound environmental deprivation early in development
    would influence the response of rhesus monkeys to a
    subsequent period of deprivation. Subjects with a
    history of previous vertical chamber confinement were
    tested 1½ years later and compared with control subjects
    with no prior history of environmental deprivation.
    Monkeys with a history of early chamber confinement
    exhibited residual effects 1½ years after their initial
    experience but before reentering the chamber. Further-
    more, the current chambering experience had a significant
    impact in terms of producing decreased environmental
    activity and increased passivity. Effects of the
    current chambering experience occurred whether or not
    individual subjects had had a prior experience. These
    data contrast with previous studies which indicate prior
    separation to be a high risk-predisposing event in
    rhesus monkeys.

391. McKinney, W.T. Jr., Young, L.D., Suomi, S.J., & Davis, J.M.
Chlorpromazine treatment of disturbed monkeys. Archives of
General Psychiatry, 1973, 29, 490-494.

    Four rhesus monkeys which were subjected to partial social
    isolation during the first 11 months of life and
    which had consistently exhibited patterns of grossly
    abnormal behavior during the 1½ years prior to the current
    study were treated with chlorpromazine. Three of the
    four subjects showed significant decreases in their
    self-disturbance behaviors on 7.5 mg/day of chlorpromazine
    given as the liquid concentrate by nasogastric
    intubation. The fourth subject showed no change. Plasma
    chlorpromazine levels confirmed absorption of the drug
    in all subjects.

392. McKinney, W.T. Jr., Eising, R.G., Moran E.C., Suomi, S.J., &
Harlow, H.F. Effects of reserpine on the social behavior of
rhesus monkeys. Diseases of the Nervous System, 1971, 32(11),
735-741.

    Reserpine was administered daily by intubation for
    81 days to three rhesus monkeys. Their behavior during
    the experimental period was compared to their behavior
    before and after the drug period, as well as to that of
    a control group of three monkeys given water instead
    of reserpine. Experimental findings were as follows:
    (1) Resperpine caused significant behavioral changes
    in the rhesus monkey. These changes included decreases
    in visual exploration and locomotion, and increases
    in self-huddling, posturing, and tremor. (2) The behavioral
    effects of repeated daily dosage were not cumulative nor
    was a tolerance developed by the Ss.

393. Meier, G.W. Behavior of infant monkeys: Differences
attributable to mode of birth. Science, 1964, 143, 968-970.

    A comparison of behavior of infant monkeys shortly
    after birth reveals differences in reactivity which can
    be related to the route of delivery, whether vaginal or
    cesarian secion. The depression of behavior in the
    surgically delivered infants persists through day
    5 postpartum and ultimately appears as lowered condi-
    tionability. Anesthetics or other drugs are not causal
    factors.

394. Meier, G.W. Maternal behavior of feral and laboratory-reared
monkeys following the surgical delivery of their infants. Nature,
1965, 206, 492-493.

   Two groups of female rhesus monkeys (Macaca mulatta), one
   feral born and the other laboratory born and individually
   reared in wire cages (without peers or surrogates, with
   only auditory and visual contact with other monkeys),
   delivered their offspring via caesarean section. The
   maternal behavior of these animals was observed and
   compared for behavioral deficits.

395. Meier, G.W. Other data on the effects of social isolation
during rearing upon adult reproductive behavior in the rhesus
monkey. Animal Behavior, 1965, 13(2-3), 228-231.

   Data are reported on the reproductive behaviours of
   14 rhesus monkeys which had been reared from infancy
   in relative social isolation. The 10 females were
   able to conceive; the 5 permitted to bear their
   young vaginally at term reacted to the offspring
   normally. The 4 males mated satisfactorily with
   species-typical positioning and, in 3 cases, sired
   offspring. The differences between these behaviours
   and those reported elsewhere are striking. The
   characteristics of the rearing conditions suggest
   the importance of visual and auditory contact in
   infancy and adolescence for the development of
   normal sexual behaviour.

396. Meier, G.W. Comments on Sackett's "innate mechanisms,
rearing conditions, and a theory of early experience effects in
primates". In M.R. Jones (Ed.), The prediction of behavior:
Early experience variables. Coral Gables, Fla.: University of
Miami Press, 1970, 55-60.

   This paper critizes Sackett's explanation of the
   failure of isolation-reared animals to respond
   effectively under subsequent examination: namely,
   that the animal reared in isolation (the rhesus
   macaque, in this instance) experiences a paucity
   of environmental feedback and thereby fails to
   develop the inhibition mode of responding necessary
   for rapid adaptation to the later testing situation.
   The author raises questions regarding the positive

features of the rearing experiences of the isolated
infant and the temporal characteristics of early
learning, both of which deal with the behavioral
content of the rearing period.

397. Meier, G.W. Toward a strategy of analysis of behavioral
development. In D. Purpura & G.P. Reaser (Eds.), Evaluation of
methodological approaches to the study of brain maturation and its
disturbances. Baltimore, Maryland: University Park, 1974, 131-
134.

The article presents a very general analysis of
organismal development under various environmental
conditions. Independent and dependent variables of
behavioral development methodologies are discussed.

398. Meier, G.W., & Berger, R.J. Development of sleep and wakefulness patterns in the infant rhesus monkey. Experimental
Neurology, 1965, 12(3), 257-277.

The sleep and wakefulness patterns of five neonatal
rhesus monkeys were monitored by continuous polygraphic
recording and compared with the animal's behavior until
the infants were 1 month old. Similar observations and
recordings were made for comparative purposes on three
juveniles of 9-13 months of age. The three stages of
vigilance; wakefulness; high voltage, slow wave sleep
(HVS); and low voltage, fast wave sleep (LVF) exist in
the same form as characterized by the established polygraphic and behavioral patterns of the adult, although
in varying temporal relations. For example, the rise
from birth in the total sleep time and that proportion
of it occupied by LVF persists until the seventh day of
life. Thereafter, the total sleep time remains constant
while the proportion of LVF declines into juvenile life.
In addition, the mean durations of periods of wakefulness,
sleep, HVS, and LVF show marked changes on this day. The
uniqueness of the seventh day of postnatal life for the
development of the infant is further emphasized by the
initial appearance at this age of a number of important
behavioral responses.

399. Meier, G.W., & Berger, R.J. Thresholds to arousing stimulation in the developing infant rhesus monkey. Psychonomic Science,
1967, 7(7), 247-248.

    Thresholds of vocalization to electroshock were
    determined in infant and juvenile monkeys during random
    epochs of wakefulness, during high-voltage, slow-
    wave sleep, and during low-voltage, fast-wave sleep.
    These thresholds increased in all states over the first
    days of postnatal life and then decreased differentially
    for each of the states thereafter.

400. Meier, G.W., & Garcia-Rodriquez, C. Continuing behavioral
differences in infant monkeys as related to mode of delivery.
Psychological Reports, 1966, 19, 1219-1225.

    The previously-described differences in CAR and CER
    acquisition during the neonatal period related
    to mode of delivery persist for at least 3 months
    thereafter providing the infant is first given the
    conditioning experience during the neonatal period.
    In the present study groups of rhesus monkeys
    (Macaca mulatta), half delivered by cesarean section
    and half delivered vaginally, were tested at monthly
    intervals starting on Day 1 or Day 31. The striking
    differences characteristic of the conditioning
    performances at birth were greatly reduced at Day 31
    in those groups first observed at that age. Although
    the age-at-introduction X mode-of-delivery interaction
    was impressive, a convincing basis for interpretation
    is not apparent. Activity and vocalization levels
    could not be related to those conditioning performances.

401. Meier, G.W., & Garcia-Rodriquez, C. Development of conditioned behaviors in the infant rhesus monkey. Psychological Reports,
1966, 19, 1159-1169.

    Infant monkeys were placed on a conditioned
    avoidance (CAR), conditioned emotion (CER),
    or pseudo-conditioned avoidance response (Pseudo-
    CAR) schedule. The CAR Ss were started on Day 1, Day 31,
    or Day 61; the CER Ss, at Day 1 or Day 31; and
    the Pseudo-CAR Ss, at Day 1. All were reintroduced
    into the situation for 5-day sessions at 30 day
    intervals until 90 days of age. Relevant comparisons

of the groups indicated that: (1) CAR was readily
established in the newborn; (2) the efficiency of CAR
was cumulative and dependent upon experience alone;
(3) the efficiency of CER was age-dependent with a
maximum at 60 or 90 days of age; (4) the establishment
of CAR does not depend upon previously established
CER.

402. Meier, G.W., & Schutzman, L.H. Mother-infant interactions
and experimental manipulation: Confounding or misidentification?
Developmental Psychobiology, 1968, 1(2), 141-145.

The role of maternal responsiveness to offspring behaviors
in the research on early experience and behavioral
development is discussed. The proposition is
offered that the behavioral changes following experimenter
manipulation of the offspring during the preweaning
period, as by handling, isolation, shock stimulation, or
cold stress, are attributable to the altered interaction
between mother and infant, such that the mother responds
differentially to certain behaviors of her offspring or
predominantly to those offspring showing such behaviors.
In support of this proposition, data are cited on
preferential maternal responsiveness to infants by sex
and the frequency is enumerated of significant
Sex X Treatment interactions in the early experience
studies.

403. Menzel, E.W. Jr. The effects of stimulus size and proximity
upon avoidance of complex objects in rhesus monkeys. Journal of
Comparative and Physiological Psychology, 1962, 55(6), 1044 -
1046.

This experiment analyzed the effects of stimulus size
and proximity upon avoidance responses in the rhesus
monkey. Three sizes of complex representational objects
were used, and objects were placed inside S's home cage
or just outside the cage. Avoidance increased directly
with object size and proximity. Familiarity with the
objects (either through repeated tests or through pretest
exposure of the stimuli) led to virtually complete
disappearance of avoidance and gave rise to object
contact. The "fear Stimuli" were destroyed by the Ss
on the eighth day of the experiment.

404. Menzel, E.W. Jr. Individual difference in the responsiveness
of young chimpanzees to stimulus size and novelty. Perceptual
and Motor Skills, 1962, 15, 127-134.

    This paper analyzed the interactions between stimulus size,
    stimulus novelty, and individual differences in
    the responses of five young chimpanzees to objects.
    Test procedure consisted of placing into S's home cage
    a single piece of wood, which ranged from 1 sq. in. to
    251 sq. in. There appeared to be an optimum size of
    objects for varied manipulatory reactions, smaller objects
    being handled in cursory fashion, and larger objects
    being contacted tentatively or not at all. The precise
    size of the optimum stimulus varied for particular
    responses, for different Ss, and also in the same S as
    a function of experience with the same or similar
    objects. Experience tended to increase the size of
    object required to produce either avoidance or vigorous
    contact activities; consequently, the behavior of
    cautious animals came over a period of weeks to resemble
    the initial behavior of bolder animals.

405. Menzel, E.W. Jr. The effects of cumulative experience on
responses to novel objects in young isolation-reared chimpanzees.
Behavior, 1963, 21, 1-12.

    Reactions toward small novel objects were studied in
    two highly cautious isolation-reared chimpanzees.
    In Experiment I the subjects were presented with 75
    different objects. Grasping increased from near-zero-
    level to high levels, and stereotyped, autistic
    activities decreased over the 25 test days. At the
    close of the experiment the subjects spent most of their
    time in object-contact. This transition from caution to
    play was more rapid for objects preexposed 16 hrs vs
    completely novel objects, and for objects constructed of
    wood vs objects constructed of other materials. Experiment
    2 tested whether responsiveness would now be maintained
    at different levels by different classes of stimuli. It
    was found that the subjects would grasp a stuffed toy
    almost incessantly and stereotype infrequently; but
    with a wood block, grasping decreased and stereotypy
    increased across several daily trials. The latter
    decrease of grasping and increase of sterotypy can be
    interpreted as a "satiation effect". Experiment 3 tested
    whether specific types of manipulatory reaction would
    now vary as a function of specific objects; the familiar

stuffed toy and wood block were contrasted with three
other, novel objects. It was found that behaviour
patterns were somewhat different for each object. Over
all three experiments, grasping and stereotypy were
reciprocally related to each other. With simple
objects such as wood blocks, both measures appeared
to be curvilinear functions of experience.

406. Menzel, E.W., Jr. Patterns of responsiveness in chimpanzees
reared through infancy under conditions of environmental restriction.
Psychologische Forschung, 1964, 27, 337-365.

The behavior of infant chimpanzees reared from
birth to 21 months under conditions of extreme
environmental and social restriction was described
and catalogued. The effects of experience upon
responses were outlined: (a) Kaspar Hausers are
similar to most chimpanzee infants raised in the laboratory
nursery. (b) The major differences at 2 years of age
are between nursery reared and mother-reared or
feral chimpanzees; and these differences tend to
decrease as a function of cumulative experience. (c)
Individual differences reside principally in the
patterning of motor behavior and in the adequate
stimuli for the release of a pattern. Descriptions
of individual differences in terms of "social" or
"emotional" are possible. The effects of stimulus
factors upon the responsiveness of restricted chimpanzees
were outlined: (a) At 2 years of age stimulus novelty
is prepotent, and novelty produces avoidance. (b)
Intensity and novelty function in similar fashions;
large amounts produce avoidance and smaller amounts
produce approach. (c) Nonintensive stimulus factors
become important principally after considerable adaptation
has occurred. It was argued that all specific drive
behaviors in physically intact infant chimpanzees can
be analyzed as facets of "responsiveness". Using the
Kaspar Hauser data as a starting point, a wide range of
chimpanzee responses was classified in terms of a single
frame of reference. (*Note: Kaspar Hauser refers to
totally isolated chimpanzees)

407. Menzel, E.W. Jr. Primate naturalistic research and problems
of early experience. Developmental Psychobioloby, 1968, 1(3),
175-184.

This paper attempts to approach behavioral development
and early experience from a naturalistic, ecologically
oriented point of view, and it reviews illustrative data
from recent primate field studies, particularly in the
area of social behavior. What is the ecological-social
context into which the infant primate must fit himself
and what are the end-points toward which normal development must move? Where does the infant locate himself
within his context? Where will he be and what will he do
at successive stages in development? These are the
sorts of questions that are posed. In general, a
naturalistic approach to development starts from that
which is already there--i.e., groups of animals in their
native habitats--and it tries to describe and analyze,
first in broad outline, and then with increasing detail,
how they come to be as they are. It supplements, comple-
ments, and in some respects reverses the traditional
approach of experimental psychology, which takes a "naive"
individual alone in an "absolutely controlled" and
usually empty environment as a theoretical model, and
tries to work from this ground up to normal ecosystems
by adding variables one at a time.

408. Menzel, E.W., Jr. Naturalistic and experimental approaches
to primate behavior. In E. Willems and H. Raush (Eds.), Naturalistic viewpoints in psychological research. New York: Holt, Rinehart
and Winston, 1969, 78-121.

Through a review of the author's own research, this paper
argues that experimental and naturalistic approaches to
primate behavior not only can be, but are, intimately
related and necessary to each other. A summary of what
naturalistic and experimental approaches have in common
is presented.

409. Menzel, E.W. Jr., Davenport, R.K. Jr., & Rogers, C.M. Some
aspects of behavior toward novelty in young chimpanzees. Journal
of Comparative and Physiological Psychology, 1961, 54(1), 16-19.

Behavior toward novelty was studied in two young
isolation-reared chimpanzees. In one experiment, Ss
were thoroughly adapted to a single object. Subsequently
they were exposed to objects that varied systematically
from the standard. Contact-duration scores were found
to vary as a function of specified objects. New objects

elicited more contact than the familiar object;
new cues elicited more contact than cues embodied in
the standard object; and objects new in two or three
cues elicited more contact than objects new in only
one cue. There was a suggestion, however, that such
preferences were not immediate. Initially, Ss appeared
cautious, any contact scores increased day by day. In
a second experiment Ss were presented a series of
"completely novel" stimulus objects and allowed to adapt
to each in turn. Early in the series they appeared
fearful or cautious, and extensive object-contact
came only after several 5-min. exposures to each
object. However, they seemed to "learn to play,"
for later in the series contacts came to be
immediate and persistent even on Trial 1.

410. Menzel, E.W. Jr., Davenport, R.K. Jr., & Rogers, C.M. Effects
of environmental restriction upon the chimpanzee's responsiveness
in novel situations. Journal of Comparative and Physiological
Psychology, 1963, 56(2), 329-334.

Two groups of young chimpanzees were captured in the field (Ns
= 3, 11) and 4 groups reared from birth in small
cubicles (Ns = 5, 4, 4, 3) were observed for 2 hr.
alone in an unfamiliar room. Eighteen of the same Ss were
tested in further sessions. Wild-born Ss differed
from restricted Ss in postural behavior, locomotion,
climbing, manipulation, self-directed responses,
vocalization, and the patterning and temporal trend
of activity. The restricted groups differed from
each other in only minor respects, and the wild-born
groups seemed different in no respects.

411. Menzel, E.W. Jr., Davenport, R.K. Jr., & Rogers, C.M. The
effects of environmental restriction upon the chimpanzee's
responsiveness to objects. Journal of Comparative and Physiological
Psychology, 1963, 56(1), 78-85.

In sharp contrast to 2-yr.-old wild-born chimpanzees,
similarly aged Ss raised in restricted environments were
timid of objects, contacted them rarely, and spent
most of their time in stereotyped self-directed
activities. Restricted Ss raised in pairs were the most
timid of all and dependent upon each other for
adaptation to stimuli. A restricted group that had had

special manipulatory experience surpassed other
isolates in object-contact, but only in situations
similar to rearing. A restricted group with
special visual experience did not differ from
maximally restricted Ss who had been reared for
21 mo. in bare gray cubicles. It was argued that
restricted chimpanzees are retarded but
potentially typical in their responsiveness
to objects.

412. Menzel, E.W. Jr., Davenport, R.K. Jr., & Rogers, C.M. The
development of tool using in wild-born and restriction-reared
chimpanzees. Folia Primatologica, 1970, 12, 273-283.

Eight near-adult chimpanzees that had been reared
for the first 2 years of life in highly restricted
laboratory environments were compared on a KÖHLER-
type stick problem with 10 wild-born controls
that had been captured well before the age at which
complex manipulative activities occur with any
reliability in the wild. For several years before
the test the two groups had shared the same social,
object, and test opportunities and all animals had
manipulated objects freely and persistently. In
the tool using, test, however, group differences
were found at every 'level' of performance except
simple stick play. General manipulatory drive,
matured motor patterns, and operant conditioning are
no doubt necessary in certain aspects of performance;
but, contrary to what authors such as HALL, HARLOW and
SCHILLER have implied, they have not been proved
sufficient to account for the 'intelligent' use of
objects in primates; and they are themselves dependent
upon early infantile experience.

413. Menzel, E.W. Jr., Davenport, R.K. Jr., & Rogers, C.M.
Protocultural aspects of chimpanzees' responsiveness to novel
objects. Folia Primatologica, 1972, 17, 161-170.

Nineteen juvenile chimpanzees, most of them with
extremely limited previous social and object
experience were housed together in 17 successive
groupings of 3 animals; the 1st trio consisted of
individuals A, B and C; the 2nd of B, C and D; the
3rd of C, D and E, and so on. Each such 'social

generation' was tested with the same two toys.
Initially both objects were avoided or approached
with caution. In the 3rd trio, however, individual
E habituated to one object and in the 5th trio
individual F habituated to the other object. An
enduring tradition of boldness and play then ensued
for each object separately. The traditions were more
stable across groups as a whole than within individuals.

414. Meyer, J.S., & Bowman, R.E. Rearing experience, stress, and
adrenocorticosteroids in the rhesus monkey. Physiology and
Behavior, 1972, 8, 339-343.

Baseline cortisol levels and adrenocortical response
to ACTH administration and chair-restraint stress
were measured in rhesus monkeys reared under partial
social isolation, total social isolation, and jungle
conditions, 3-4 years after the rearing treatments. In
contrast to expectations derived from the rodent
literature on early treatment effects, no pituitary-
adrenocortical effects of rearing condition were found
in the monkey. From these and other data, it seems
that such effects which may appear shortly after rearing
treatment in the monkey are transient and there is no
long-term modification of a hormonostat mechanism
regulating basal cortisol levels or adrenocortical
responsivity. Other effects of stress were observed;
additionally, some sex differences were noted. Adrenocortical secretion apparently fell to zero soon after
termination of stress and did not recommence within the
next 2 hr, indicating a rapid and complete poststress
reset of some feedback mechanism for control of plasma
cortisol levels. Finally, plasma levels of injected,
radioactive cortisol were decreased during stress, but
recovered poststress, compared with baseline conditions,
suggesting a stress-induced shift in cortisol compartment-
alization ratios.

415. Meyer, J.S., Novak, M.A., Bowman, R.E., & Harlow, H.F.
Behavioral and hormonal effects of attachment object separation
in surrogate-peer-reared and mother-reared infant rhesus monkeys.
Developmental Psychobiology, 1975, 8(5), 425-435.

Mother-reared and surrogate-peer-reared rhesus
monkeys were separated from their respective
attachment objects at 6 months of age and tested

for the following 9 weeks to determine their
home-cage behavior and their pituitary-adrenocortical
responses to stress. Both groups displayed a
strong immediate behavioral response to separation
which was characterized by increased vocalization,
increased locomotion, and decreased self-play.
However, the surrogate-peer-reared infants showed
a subsequent recovery in their levels of self-play
whereas the mother-reared infants instead developed
stereotypic behavior patterns such as repetitive
pacing. The 2 groups displayed similar plasma
cortisol responses to weekly sessions in an apparatus
equipped with animated toy "monsters." Mother-reared
but not surrogate-peer-reared subjects, however, also
manifested elevated cortisol levels when an animal
in an adjacent cage was captured and removed for
stress testing. Mother-reared infant monkeys thus
responded in a stronger and more prolonged manner
to the loss of their attachment object than surrogate-
peer-reared infants. These results suggest that
infant rhesus monkeys form stronger attachments to
monkey mothers than to inanimate surrogate mothers,
a phenomenon which has not been as clearly demonstrated
using other indices of attachment strength.

416. Miller, R.E., Caul, W.F., & Mersky, I.A. Communication of
affects between feral and socially isolated monkeys. Journal of
Personality and Social Psychology, 1967, 7(3), 231-239.

Three rhesus monkeys were obtained from the Regional Primate
Center, University of Wisconsin. These animals had
been subjected to total social isolation during their
1st yr. of life. Three feral monkeys of the same age
constituted the normal group. The animals were
trained to perform an instrumental avoidance response
to a visual stimulus. There were no differences in
acquisition of the response either in terms of
instrumental behavior or conditioned cardiac responses.
The animals were then paired in all possible combinations
for communication of affects tests using the cooperative-
avoidance technique. The monkey which received the
conditioned stimulus was visible to the animal having
access to the response bar via closed-circuit television.
The results, both instrumental and physiological, indicated
that isolate monkeys were incapable of utilizing
facial expressions of other monkeys in order to perform
appropriate avoidance responses. The isolates also were

found to be defective senders of facial expression.

417. Miller, R.E., Caul, W.F., & Mirsky, I.A. Patterns of eating and drinking in socially-isolated rhesus monkeys. Physiology and Behavior, 1971, 7, 127-134.

   Three normally-reared and three animals which had
   been totally-isolated from social contact for the
   first year of life were tested in a series of
   experiments to determine the pattern and rate
   of eating and drinking. The monkeys were nine
   years old during these experiments and had been
   subjected to a variety of social and experimental
   conditions post-isolation. While the social isolates
   did not differ from the controls in the frequency
   and scheduling of eating and drinking during a day,
   they did consume more fluid and food during each
   meal. The isolates ate approximately 30 per cent
   more food per day over the course of a six month
   ad lib feeding period. Nevertheless, after an
   initial gain, they did not gain significantly
   more weight than did the controls. There was no
   increased motor activity among the isolates.

418. Miller, R.E., Mirsky, I.A., Caul, W.F., Sakata, T.
Hyperphagia and polydipsia in socially isolated rhesus monkeys.
Science, 1969, 165, 1027-1028.

   Three rhesus monkeys which had been isolated from
   social contact during their first year of life
   persistently overate and overdrank during
   adulthood. These monkeys ingested approximately
   twice as much fluid and food as the control animals
   reared normally.

419. Miminoshvili, D.I. Experimental neurosis in monkeys. In
I.A. Rurkin (Ed.), Theoretical and practical problems of medical
and biology experiments. New York: Pergamon Press, 1960, 53-67.

   Five monkeys were utilized for the experiment: 2
   rhesus macaques and 3 sacred baboons. Two of the
   monkeys were placed in a special stand during the
   experiments, the remaining animals were allowed to
   move freely about the dwelling cage. The purpose

of the research investigation was to examine the effect
of multiple excitatory processes in the etiology of higher
nervous system breakdown.

420. Minami, T. Early mother-infant relations in Japanese
monkeys. In Contemporary Primatology, 5th International Congress
of Primatology, Nagoya, 1974, 334-340, (Karger, Basel, 1975).

This investigation was conducted in order to clarify the
relationship between patterns of stereotyped behaviors
and infant development. Four male infants and their
mothers were observed during the first 9 months of the
infant's life. These data were compared with infants
reared in isolation. It is suggested that the patterns
of stereotyped behaviors appear in connection with
stages of development of the infant.

421. Mineka, S., & Suomi, S.J. Social separation in monkeys.
Psychological Bulletin, 1978, 85(6), 1376-1400.

Phenomena associated with social separation from
attachment objects in nonhuman primates are
comprehensively reviewed. A biphasic protest-despair
reaction to social separation is often seen in monkeys,
as in human children. However, upon reunion there is
generally a temporary increase in attachment behaviors
rather than a temporary phase of detachment as has been
reported in the human literature. Gross factors such as
age and sex do not appear to influence the responses
to separation or reunion substantially. Rather, behavioral
repertoires prior to separation and the nature of the
separation and reunion environments appear to be more
important determinants of the severity of separation
reactions. These findings are consistent with the
human literature. The evidence bearing on possible
long-term consequences of early separations is also
discussed. Finally, four different theoretical
treatments of separation phenomena are presented and
evaluated in light of existing data: Bowlby's
attachment-object-loss theory, Kaufman's conservation-
withdrawal theory, Seligman's learned helplessness
theory, and Solomon and Corbit's opponent-process
theory.

422. Missakian, E.A. Reproductive behavior of socially deprived
male rhesus monkeys. 2nd International Congress of Primatology,
Atlanta, Ga., 1968. (Abstract)

> Observations were made of the reproductive behavior of
> three groups of male rhesus monkeys (Macaca mulatta):
> wild-reared, cage-reared with adult social experience
> (periodic exposure to receptive females) and naive
> cage-reared. Results from this study supported Mason (1960)
> and Harlow's (1965) findings that early social
> deprivation produces deficits in adult social behavior.
> The extent and severity of the deficits were reflected in
> the failure of any cage-reared male to execute a normal
> mount. There was no difference between the experienced
> and naive cage-reared males on measures of frequency and
> duration of male and female grooming, grooming demand,
> threat or mount attempt. This finding suggests that the
> deficits were not reversed in males who had had periodic
> social experience with females. All measures of social
> behavior revealed that the behavior of both groups of
> cage-reared males was consistently inferior to that
> of wild-reared monkeys. Thus, the effects of social
> deprivation were not restricted to reproductive
> activity, but were extensive and generalized across
> other patterns of social behavior.

423. Missakian, E.A. Reproductive behavior of socially deprived
male rhesus monkeys (Macaca mulatta). Journal of Comparative and
Physiological Psychology, 1969, 69(3), 403-407.

> Behavioral observations were made of three groups of
> male rhesus monkeys: wild-reared, socially
> deprived with adult social experience (periodic
> exposure to sexually receptive females), and naive
> socially deprived males. Results of the observations
> supported previous conclusions regarding the effects
> of social deprivation on patterns of adult reproductive
> behavior. The extent of the deficit produced by social
> deprivation was reflected in the finding that no
> socially deprived male, experienced or naive, executed
> and appropriately oriented mount. The effects of rearing
> under conditions of social deprivation generalized across
> the other behavior categories of grooming frequency and
> duration, threat, grooming demand, and mount attempt.
> The absence of a difference, on any behavior category
> between experienced vs. naive socially deprived males
> suggested that the effects of deprivation could not

be reversed by the duration and type of experience
afforded.

424. Missakian, E.A. Effects of adult social experiences on patterns of reproductive activity of socially deprived male rhesus
monkeys (Macaca mulatta). Journal of Personality and Social Psychology, 1972, 21(1), 131-134.

The problem under investigation involved the extent to
which social group experience as an adult could modify
and/or reverse atypical behaviors produced by rearing under
conditions of social deprivation. Three socially deprived
adult male rhesus monkeys were provided with 3-6 months
experience in social groups containing emotional rather
than chronological peers. The group experience modified
the social behavior of two of the three subjects. The
nature of the changes involved an increase in social
grooming, a decrease in physical aggression, and an increase
in mounting activity. The increase in these behaviors was
accompanied by decreases both in steretoped locomotion
and self aggression. The failure of this social experience
to produce change in one of the subjects provided
valuable information about possible factors involved in
the behavioral change. Perhaps the most important
difference was the failure of this male to engage in social
play while in the group. Data from this study indicate a
direction for future research regarding the question of
reversibility of the effects of social deprivation. The
type and duration of social experience afforded these
monkeys represents the only instance of behavior modification
in socially deprived rhesus monkeys.

425. Mitchell, G. Attachment differences in male and female infant
monkeys. Child Development, 1968, 39, 611-620.

The mother's relation with the male infant was compared
with the mother's relation with the female infant in
32 mother-infant pairs of rhesus monkeys. Mothers had
more physical contact with female infants and restrained
females more frequently than males. Mothers of males
withdrew from, played with, and presented to their infants
more often than did mothers of females. Males bit their
mothers more often than did females. The frequency and

form of mother-infant contacts depend on the
behavior of the mother and the age and sex
of the infant.

426. Mitchell, G.D. Persistent behavior pathology in rhesus
monkeys following early social isolation. Folia Primatologica,
1968, 8, 132-147.

Isolate monkeys show more emotional behavior than
controls but less play, sex, exploration and
vocalization. Isolation from months 6 to 12
produces more exploration than isolation from
months 0 to 12 or 0 to 6. The isolates' hostility
toward infants and overall fear decline with age,
but hostility toward age-mates, sex, and exploration
increase with age. Infants elicit both play and
hostility from isolates. Pacing and rocking are
inversely related to bizarre idiosyncratic movements.

427. Mitchell, G.E. Abnormal behavior in primates. In L.A.
Rosenblum (Ed.), Primate behavior: Developments in field and
laboratory research (Vol. 1). Glenview, Illinois: Scott,
Foresman, & Comapny, 1972, 195-249.

This extensive review article presents a thorough
analysis of abnormal behavior in primates. Various
developmental factors as well as abnormalities related
to birth involve causative factors such as asphyxia
neonatorum, age, parity of the mother, mode of birth,
length of labor, and difficulty of labor. The effects
of early social experience such as peer deprivation,
maternal deprivation, temporary maternal separation,
and social isolation are discussed with regard to
psychopathology in nonhuman primates.

428. Mitchell, G. Comparative development of social and emotional
behavior. In G. Bermant (Ed.), Perspectives on animal behavior.
Glenview, Ill.: Scott, Foresman & Company, 1972.

To illustrate the approach of comparative developmental
psychology, this chapter explores a single topic in
this field--the development of social behavior and
social emotions. By studying the way in which social

behavior develops in other species, as well as our
own, the author feels we can better understand how
and why humans form emotional attachements to their
mothers, come to love or hate the people around them,
and gradually acquire the spectrum of abilities which
make up being a social creature. This chapter is
intended to give the student a selected veiw of
contemporary ideas and research on this topic.

429. Mitchell, G. Looking behavior in the rhesus monkey. Journal
of Phenomenlogical Psychology, 1972, 3(1), 53-67.

This report reviews information on the directions,
frequencies, and durations of looking behavior in
rhesus monkeys (Macaca mulatta). An attempt is made
to relate this information on looking to emotional
behavior and to the normal and abnormal psychological
development of the individual.

430. Mitchell, G. Syndromes resulting from social isolation of
primates. In J.H. Cullen (Ed.), Experimental behavior: A basis
for the study of mental disturbance. Dublin: Irish University
Press, 1974, 216-223.

This paper summarizes several projects which deal
with the development of disorganization in the
behavior of the rhesus monkey (Macaca mulatta).
Postural communication and bodily movement, vocalizations,
facial expressions, object of the display, and therapy
for adult male isolate-reared monkeys are discussed.

431. Mitchell, G.D. What monkeys can tell us about human violence.
The Futurist, 1975, 9(2), 75-80.

This article discusses experiments with nonhuman
primates subjected to early social deprivation.
Results are presented which support theories of
early deprivation as a cause of violence: monkeys
deprived of physical contact with other monkeys from
birth grow up to be aggressive, fearful, sexually
abnormal adults. Although the effects of severe social
deprivation may never be totally erased, the passage of
time and interaction with younger animals and peers may
make an isolate less socially maladroit and even

capable of giving and receiving affection. This
could be of importance in treating pathology in
adult humans, and in suggesting ways to forestall
the destructive acts of violence-prone adults who
were affection-starved and socially deprived as
children.

432. Mitchell, G., & Brandt, E.M. Behavioral differences related
to experience of mother and sex of infant in the rhesus monkey.
Developmental Psychology, 1970, 3(1), 149.

The Mitchell and Stevens (1968), Macaca mulatta mothers
and infants, studied from 1 to 3 months, were used in
the present study. The effects of maternal experience
and the sex of the infant on maternal and infant
behavior were evaluated again in the second 3 months.
The results suggest that the factors of maternal
experience and sex of infant do affect the behaviors
of mother and infant rhesus monkeys. In the second
3 months, maternal experience wanes in importance as
a factor influencing maternal or infant behavior, as the
sex of the infant becomes the more important factor.
One can best characterize mothers of males as "punishers,"
mothers of females as "protectors," male infants as
"doers," and female infants as "watchers." The
mother plays a role in prompting the greater
independence and activity that is typical of males.

433. Mitchell, G.D., & Clark, D.L. Long-term effects of
social isolation in nonsocially adapted rhesus monkeys. Journal
of Genetic Psychology, 1968, 113, 117-128.

Four surrogate reared monkeys, socially isolated
between three and nine months of age and who had
been well adapted to the postisolation test room
were compared with four similarly raised nonisolated
controls and with four feral mothered Ss. Subjects
from all three groups were tested at approximately
18 months of age in a nonsocial situation and in a
series of social pairings with three stimulus
strangers of equal age. It was concluded:
(1) Monkeys raised without a real mother and with
only 140 to 147 hours of peer experience during the
first year did not redirect their hostility from a
species-mate in an appropriate manner. (2) Total

social isolation led to a restricted ability to form
normal dominance relationships with social partners.
The isolates were immature in their play behavior and
emotional in social interactions. (3) Continuing
adaptation of the isolates to the postisolation test
room lowered the level of nonsocial disturbance. These
animals manually explored more than the nonisolates.
There was no evidence of an emergence phenomenon.
(4) Monkeys who received relatively little social
experience during the first year displayed more
disturbance behaviors, such as bizarre movements,
crouching, rocking, and digit sucking, when compared
to real mother-peer reared animals. They also exhibited
social hostility and assertive play more frequently
than did the more adequately socialized animals, but
showed infrequent and inappropriate sexual behavior.

434. Mitchell, G., & Redican, W. Communication in normal and
abnormal rhesus monkeys. The 29th International Congress of
Psychology. Tokyo: Sasaki Co., 1972, 237.

Isolate-reared rhesus monkeys of all ages (Macaca
mulatta) are compared with socially-reared monkeys
with regard to postural communication and bodily
movements, vocalizations, facial expressions, looking
behavior and self-directed behavior. The movements
of isolates are slow and awkward; young isolates
crouch and cower; isolates of all ages show abnormal
sexual posturing. Stereotyped and bizarre movements
also change communication. Isolates also misuse
vocalizations and facial expressions, but it is
their abnormalities in looking behavior and in
knowing self from not self which probably contribute
most to their failure to communicate adequately.

435. Mitchell, G., & Stevens, C.W. Primiparous and multiparous
monkey mothers in a mildly stressful social situation: First
three months. Developmental Psychobiology, 1968, 1(4), 280-286.

Eight primiparous rhesus monkey mothers were
matched with 8 multiparous rhesus monkey mothers
with regard to date of delivery and sex of infant.
Each mother was housed and tested individually with
her infant to preclude the compensating effects of
peer experience. The test situation involved transporting

each mother-infant pair from the home cage to a test
cage of similar size where the pair was visually
exposed to a stange mother-infant pair and human
observers. The primiparous mothers looked at, threatened, fear grimaced, and lip-smacked to these social
stimuli significantly more frequently than did the multiparous females. In addition, the inexperienced mothers
stroked or petted their infants significantly more frequently than the experienced mothers. The results support
the idea that primiparous mothers are more "anxious" or
concerned for their infants' welfare than are multiparous
mothers.

436. Mitchell, G.D., Arling, G.L., & Moller, G.W. Long-term effects
of maternal punishment on the behavior of monkeys. Psychonomic
Science, 1967, 8(5), 209-210.

Adolescent monkeys having punitive mothers for the first
three months of life display more aggression and less
social exploration than adolescent monkeys having non-
punitive mothers.

437. Mitchell, G.D., Harlow, H.F., Griffin, G.A., & Moller, G.W.
Repeated maternal separation in the monkey. Psychonomic Science,
1967, 8(5), 197-198.

Repeated early maternal separations increase the levels
of coo vocalizations and fear. This syndrome persists
when the Ss are tested one year after the final maternal
separation.

438. Mitchell, G.D., Raymond, E.J., Ruppenthal, G.C., & Harlow,
H.F. Long-term effects of total social isolation upon behavior of
rhesus monkeys. Psychological Reports, 1966, 18, 567-580.

Eight isolate monkeys were compared in a follow-up
study to 8 sophisticated controls in brief cross-
sectional pairings with 12 stimulus strangers: 4 adults,
4 age-mates, and 4 juveniles. The isolates were
characterized by infantile disturbance, less
environmental orality, more fear, more aggression,
less sex, less play, and bizarre ritualistic

movements. Twelve-mo. isolates were fearful and
nonaggressive but threatened many attacks. Six-mo.
isolates were fearful and physically aggressive.
The 12-mo. isolates demonstrated practically no
positive social behavior. Conclusions are: (a)
6 mo. of social isolation during the first year
has negative effects on social behavior up to
puberty, (b) abnormal aggression appears in 3-yr.-
old 6-mo. isolates, and (c) 12 mo. of isolation
suppress or delay this aggression.

439. Mitchell, G.D., Ruppenthal, G.C., Raymond, J., & Harlow, H.F.
Long-term effects of multiparous and primiparous monkey mother
rearing. Child Development, 1966, 37, 781-791.

Four adolescent rhesus monkeys born of primiparous
mothers were compared with 4 adolescents of
multiparous mothers. Their social and nonsocial
behaviors when paired with adult, age-mate, and
juvenile strangers were recorded. The primiparous-
mothered adolescents played less with their
partners, displayed fewer spontaneous emotionality
indexes, exhibited less self-directed behavior,
and were more disturbed in the playroom. The
multiparious-mothered monkeys were observed to
explore themselves orally more frequent in the
presence of adult strangers and were more hostile
toward them. Yet, the progeny of the experienced
mothers were much more relaxed and playful,
particularly the males. It is concluded that (a)
"individual peculiarities" with regard to play
persist in the primiparous-mothered males at
puberty; (b) subtle but significant differences
between the groups appear either as a consequence of
age or of the more demanding stimulus stranger
situation; and (c) the progeny of the more rejecting
mothers (more experienced mothers) are more hostile,
but this hostility is neither brutal nor abnormal.

440. Miller, G.W., Harlow, H.F., & Mitchell, G.D. Factors
affecting agonistic communication in rhesus monkeys (Macaca
mulatta). Behavior, 1968, 31, 339-357.

This study deals with factors affecting agonistic
communication in rhesus monkeys (Macaca mulatta).

Definitions are given for various postures, facial
expressions, and vocalizations. The data revealed
that such communications depend at least upon the
following factors: (1) age; (2) sex; (3) rearing;
(4) social stimulus; and (5) adaptation to the
social partner.

441. Moran, E.C. & McKinney, W.T. Jr. Effects of chlorpromazine
on the vertical chamber syndrome in rhesus monkeys. Archives of
General Psychiatry, 1975, 32, 1409-1413.

In an attempt at social rehabilitation, chlorpromazine
was given to three groups of rhesus monkeys that
had been confined to the vertical chamber apparatus
early in their development. Previous studies have
shown that such periods of deprivation produce
severe deficits in social behavior. There were no
substantial beneficial effects of chlorpromazine
treatment; however, there was a notable amount of
spontaneous improvement seen in all three groups.
We discuss these data in terms of their implications
for the use of the vertical chamber as a tool in
experimental research of psychopathological
disorder.

442. Morris, D. The response of animals to a restricted
environment. Symposium of the Zoological Society of London, 1964,
13, 99-118.

This article presents a broad discussion of animal
behavior in sterile-restricted environments.
Specific reference is made to zoological conditions
and measures to minimize maladaptive behavior patterns.

443. Morrison, H.L., & McKinney, W.T. Jr. Environments of
dysfunction: The relevance of primate animal models. In R.N.
Walsh and W.T. Greenough,(Eds.), Environments as therapy for brain
dysfunction. New York: Plenum Press, 1976.

This chapter discusses the current status of experimental
psychopathology and concentrates on the advantages and
disadvantages, as well as the criteria for the development and assessment, of animal models. The various
social and biologic environmental induction techniques

and concomitant rehabilitative techniques of a biologic
and social nature are presented. Additionally,
correlative studies of biologic systems affected by
these techniques are included.

444. Morrison, H.L., & McKinney, W.T., Jr. Models of human psychopathology: Experimental approaches in primates. In A. Frazer
and A. Winokur (Eds.), Biological bases of psychiatric disorders.
New York: Spectrum, 1977.

This chapter focuses on the development of the field of
experimental primatology and its importance in contributing
to developmental theories of behavior. The utilization
of specific social and biological induction techniques to
produce syndromes of abnormal behavior in primates are
discussed. Areas of importance for an understanding of
the development of abnormal behavior include variables
important in the development of a reciprocal mother-infant
attachment bond, short- and long-term effects of varying
conditions of rearing on the social behavior of primates,
the role of separation in the production of abnormal
behavior on short- and long-term behavior, the interactional
nature between social and biological factors in the production of abnormal behavior, and the rehabilitative
approaches effective in ameliorating the syndromes of
abnormal behavior in primates. A summary of those concepts
most important to the field of psychiatry and a considera-
tion of perspectives for the future are included.

445. Morrison, H.L. & McKinney, W.T. Jr. Models of psychological
dysfunction: Ethological and psychiatric contributions. In M.T.
McGuire and L.A. Fairbanks (Eds.), Ethological psychiatry: Psychopathology in the context of evolutionary biology. New York: Grune
and Stratton, 1977.

This chapter reviews the foundation of experimental
animal models of neurosis and depression and outlines
the value of extending psychiatric thought to
encompass new data sets that can, in turn, aid
in evaluating existing theories and modes of thought.
The authors view behavior and psychopathology as having
multiple causes and demonstrate how an understanding of
the interrelations between variables is facilitated through
the use of primate models. Once established through
empirical checking for similarities in form and response,

the animal model can then be used to answer many of the
questions outlined in Chapter 1, particularly those
dealing with biochemical and experiential effects on
symptomatic behavior and on the stability versus
lability of specific behavioral disorders.

446. Mowbray, J.B., & Cadell, T.E. Early behavior patterns in
rhesus monkeys. Journal of Comparative and Physiological Psychol-
logy, 1962, 55(3), 350-357.

Fifteen infant monkeys were tested from birth to the
twenty-fifth day of age to determine consistency,
degree of volition involved, and mean time of appearance
and disappearance of 10 early behavior patterns related
to nursing and maintenance of contact with the mother.
Patterns are defined and the testing procedures used
for each pattern are described. Results are discussed in
terms of previous work in the area and tabulated for
use in assessing the development of rhesus neonates.

447. Nadler, R.D. Determinants of variability in maternal behavior
of captive female gorillas. In S. Kondo, M. Kawai, A. Ehara and
S. Kawamura (Eds.), Proceedings of the Symposium of The 5th
International Congress of Primatological Society, Tokyo: Japan
Science Press, 1975.

Information on more than 90 gorilla births in captivity
was analyzed in order to assess the degrees of maternal
competency of the mothers and to determine which factors
contributed to variability in maternal behavior. It
was found that more than 80% of all live-born infants
were separated from their mothers. Separation of
offspring within the first week of life for primiparous
females was associated predominantly with abuse and
neglect of infants, whereas for multipara, offspring
were separated in anticipation of injury to, or
illness of, the infants. Taken as a whole, multiparous
females exhibited more adequate maternal behavior
than did primipiara. The data futher indicated, that
14 of 16 multiparous females on whom data were available
showed improvement in their maternal proficiency over
their primiparous performance.

448.   Nadler, R.D., & Braggio, J.T.   Sex and species differences in
captive-reared juvenile chimpanzees and orang-utans.   Journal of
Human Evolution, 1974, 3, 541-550.

     The behavior of nursery-reared juvenile chimpanzees
     and orang-utans was studied during intraspecific
     group interactions in large outdoor compounds.
     Chimpanzees spent more time on the ground, more
     time in proximity to another and exhibited more
     slap-hit behavior than the orang-utans.   Both
     groups exhibited only species-typical vocalizations.
     Males of both species were more assertive in
     social interactions, whereas females interacted
     more with inanimate objects.   Positive correlations
     between body weight and the frequency of some
     social interactions partially confounded
     interpretation of the sex differences.   The results
     indicate that early experience with adults is
     not necessary for the development of certain sexually
     dimorphic and species-typical behavior patterns
     in juvenile apes.   However, early experience with peers
     seems to facilitate development of these patterns.
     The results are discussed in relation to available
     data on the early hormonal and social determinants
     of behavior in the higher

449.   Nadler, R.D., & Green, S.   Separation and reunion of a
gorilla infant and mother.   International Zoo Yearbook, 1975, 15,
198-201.

     This article describes the separation and subsequent
     reunion of a Lowland gorilla infant (Gorilla g. gorilla)
     and its mother.   The significance of the event lies
     in the fact that separation for a prolonged interval
     and reunion took place while the infant was less than
     one year old, a stage of development in gorillas that
     is characterized by a strong dependency on the mother.

450. Nadler, R.D., & Rosenblum, L.A. Factors influencing sexual
behavior of male bonnet macaques (<u>Macaca radiata</u>). In H. Kammer
(Ed.), <u>Proceedings of the 3rd International Congress of Primatology</u>,
<u>3</u>, (Karger, Basel 1971)

> Adult wild-born bonnet macaques were tested for sexual
> behavior duing 1 h exhaustion and recovery tests.
> Laboratory-born, socially-reared bonnets, 2-5 years old
> received 30 min tests. Male sexual behavior was
> influenced by the male's prior experience with the
> stimulus female, dominance vis-a-vis the female, age,
> unknown factors(s) associated with individual stimulus
> females and recent mating activity.

451. Newman, J.D. & Symmes, D. Abnormal vocalizations in
isolation-reared monkeys. <u>Proceedings of the American Psychological
Association</u>, 1971, <u>6</u>, 789. (Abstract)

> Results of spectrographic analyses are presented which
> specify the abnormal nature of vocalizations given by
> isolation-reared rhesus monkeys. Two groups of four
> monkeys were the subjects of neurobehavioral studies
> during the period 6-24 mo. of age, after differential
> rearing for the first 6 mo. of life. The experimental
> animals were separated from their mothers shortly after
> birth and reared without physical contact with other
> monkeys. The control animals were raised with their
> mothers during this time. Both groups were raised in an
> otherwise similar laboratory environment, rich in
> typical rhesus vocalizations. Vocalizations were
> recorded from the eight monkeys during observations
> spanning the 18-mo. period in the laboratory. In each
> session, the monkey was by itself in the observation
> cage but in auditory contact with other adult and
> juvenile monkeys. Television records of these sessions
> help confirm the source of each vocalization. The
> control group produced only species-typical harsh and
> clear calls in every session. The experimental group,
> on the other hand, produced vocalizations which were
> aberrant in several ways. Major abnormalities occurred
> as unusual temporal organization and abrupt elevation,
> diminution, or fluctuation in acoustic intensity level.
> Clear calls show abnormal harmonic emphasis and lack
> the inflection typical of the normal group's clear
> calls. Threatening calls, such as growls, were
> recorded from some of the experimental animals, but
> were rare. These results suggest that monkeys raised
> under conditions of maternal deprivation can develop a

highly abnormal system of auditory connumication, which
persists until at least 2 yr. of life.

452.   Newman, J.D., & Symmes, D.   Vocal pathology in socially
deprived monkeys.   Developmental Psychobiology, 1974, 7(4), 351-
358.

   Structural abnormalities were found in the clear calls
   of rhesus monkeys raised in partial social isolation.
   These abnormalities included abrupt pitch changes,
   harmonic emphasis shifts, temporal discontinuity, and
   lack of the characteristic inflection found in such
   calls from mother-reared control monkeys.   Other
   forms of vocalization appeared structurally normal.
   Vocal pathologies were distinctive for each isolate
   tested and persisted over the age range 8-24 months.

453.   Nissen, H.W., Chow, K.L., & Semmes, J.   Effects of restricted
opportunity for tactual, kinesthetic, and manipulative experience
on the behavior of a chimpanzee.   American Journal of Psychology,
1951, 64(4), 485-507.

   Tactual, kinesthetic, and manipulative experience was
   restricted in a young male chimpanzee from the age of
   1 month to 31 months by encasing the limbs in
   cardboard cylinders.   Effects of the restriction
   were assessed by (1) observations of general behavior
   exhibited during the 30 months, (2) training in
   visual discrimination of size, form and depth, (3)
   training in a tactual discrimination, (4) a series of
   tactual-motor tests (with cylinders removed) at
   ages 14 and 31 months, and (5) observations of
   behavioral changes during the 4 months following
   permanent removal of the cylinders.

454.   Noble, A.B., McKinney, W.T. Jr., Mohr, D. & Moran, E.
Diazepam treatment of socially isolated monkeys.   American Journal
of Psychiatry, 1976, 133(10), 1165-1170.

   Four rhesus monkeys were reared for the first eight
   months of life in total social isolation.   One animal
   died during this period; the three remaining subjects

were treated with diazepam in an isolation chamber, in
their home cages, and in a playroom testing situation.
Diazepam significantly decreased the self-disturbance
behaviors of two subjects, and there was even the
appearance of some social behaviors, although they
were limited and not of the same quality as in
nonisolated subjects. The authors discuss the
implications of the data for understanding the
significance of the social isolation syndrome in
monkeys as a model for human psychoses.

455. Novak, M.A. Fear-attachment relationships in infant and
juvenile rhesus monkeys (Doctoral dissertation, University of
Wisconsin, 1973). Dissertation Abstracts International, 1974, 34
(10).

This study investigated the relative capabilities of
biological mothers, age-mate infants, and inanimate
surrogates with regard to reducing the fear of infant
monkeys (Macaca mulatta) in a stressful situation.
Behavioral observations were supplemented with plasma
cortisol measurements to provide a physiological
index of stress and home cage data were obtained and
compared with the results of fear testing. It was
concluded that diverse attachment objects differ
in their fear reducing capabilities, that infant
differences in fear responsivity may not be discernable
immediately following attachment object separation due
to the reaction to separation itself, and that
familiar peers may serve a similar function in certain
juvenile monkeys depending on their early rearing
experience.

456. Novak. M.A., & Harlow, H.F. Social recovery of monkeys
isolated for the first year of life: I. Rehabilitation and
therapy. Developmental Psychology, 1975, 11(4), 453-465.

Previous research demonstrated that 12 months of total
social isolation initiated at birth produced severe
and seemingly permanent social deficits in rhesus
monkey subjects. Such monkeys exhibited self-clasping,
self-mouthing, and other stereotypic, self-directed
responses instead of the appropriate species-typical
behaviors. Although early experimentation designed to
rehabilitate isolate-reared subjects was not

successful, recent research has indicated that 6-month-
isolated monkeys could develop social behaviors if
exposed to younger, socially unsophisticated "therapist"
monkeys. The present experiment demonstrated that 12-
month isolate-reared monkeys developed appropriate
species-typical behavior through the use of adaptation,
self-pacing of visual input, and exposure to younger
"therapist" monkeys. Adaptation enabled the isolate
monkeys to become familiar with their postisolation
environment, while self-pacing facilitated their
watching the therapist monkeys' social interactions.
So primed, the isolates showed a marked decrease in
self-directed behaviors following extensive intimate
contact with the therapists. As species-typical
behaviors significantly increased during this period,
the isolate behavioral repertoire did not differ
substantially from the therapist behavioral repertoire
by the end of the therapy period. Such results clearly
fail to support a critical period for socialization
in the rhesus monkey, and an alternative environment-
specific learning hypothesis is proposed.

457. Passingham, R.E.   Studies with non-human primates.   In H.J.
Eysenck (Ed.), Handbook of abnormal psychology. New York: Basic
Books, 1973, 624-644.

A thorough review of non-human primate research is
presented with various aspects of biology, psychopathology,
social behavior, and environmental restriction being
discussed.

458. Ploog, D.   Early communication processes in squirrel monkeys.
In R.J. Robinson (Ed.), Brain and early behavior.   New York:
Academic Press, 1968.

The behavioral ontogeny and early communication processes
in infant squirrel monkeys raised in mixed groups were
studied from birth to the third year. The complete
development is divided into seven phases. From brain
stimulation experiments on adult squirrel monkeys and
from neuroanatomical data on brain maturation it can
be inferred that brain structures important for complex
communication processes are already fulfilling their
function at birth. Specific characteristics of early
social communication in primates are discussed.

459. Pratt, C.L. The developmental consequences of variations in early social stimulation (Doctoral dissertation, University of Wisconsin, 1969). Dissertation Abstracts International, 1970, 31(1).

> This study assesses the effects of rearing rhesus monkeys
> (Macaca mulatta) with different amounts of early
> social stimulation. The methodological consequences
> of the data were discussed, as were the advantages of
> presenting social data in the form of profiles of
> behavior change probabilities.

460. Pratt, C.L., & Sackett, G.P. Selection of social partners as a function of peer contact during rearing. Science, 1967, 155, 1133-1135.

> Three groups of monkeys were raised with different
> degrees of contact with their peers. The first
> group was allowed no contact, the second only
> visual and auditory contact, and the third was
> allowed complete and normal contact with their peers.
> Animals of all three groups were allowed to interact
> socially; they were then tested for their preference
> for monkeys raised under the same conditions or for
> monkeys raised under different conditions. Monkeys
> raised under the same conditions preferred each
> other, even if the stimulus animals were completely
> strange to the test monkey.

461. Prescott, J.W. Central nervous system functioning in altered sensory environment (S.I. Cohen). Invited commentary in M.H. Appley and R. Trumbull (Eds.), Psychological stress. New York: Appleton-Century Crofts, 1967, 113-120.

> This commentary takes issue with Dr. Cohen's conclusions,
> namely, that it is fallacious to think of sensory
> deprivation as a unified experimental condition and
> unreasonable to seek a single explanatory construct to
> integrate the differing experimental findings. The
> author feels that an examination of early experiences
> which modify and determine, through developmental
> processes, those neurophysiological structures that
> mediate perception, learning motivation, and affect,
> may provide such an explanatory construct. He proposes
> that sensory deprivation--or restricted experiences--

in the early critical and formative years of development results in a permanent neurophysiological deficit that manifests itself by chronic "stimulation-seeking" behavior (i.e., undue dependence upon environmental stimulation) for the maintenance of controlled and integrated behavior by the deprived organism.

462. Prescott, J.W. Early social deprivation. In D.B. Lindsley and A.H. Riesen (Eds.), Perspectives on human deprivation, biological, psychological, and sociological. Chapter IV, Biological Substrates of Development and Behavior. Bethesda, Maryland: N.I. of Child Health and Human Development, 1968, 225-256.

In this paper Prescott discusses data which provide a bridge between the social and biological disciplines through the phenomena of sensory experience and its deprivation during ontogenetic development. The possibility of identifying specific neural structures in the specification of a biological predisposition to violent-aggressive behavior and impaired socialization as a consequence of lack of early sensory-social experience, is explored.

463. Prescott, J.W. Early somatosensory deprivation as an ontogenetic process in the abnormal development of the brain and behavior. Medical Primatology 1970. Proceedings of the 2nd Conference of Experimental Medical Surgery in Primates, New York 1969, 356-375 (Karger, Basel 1971)

An outline of a developmental neural-behavioral theory of maternal-social deprivation (MSD) is presented which attempts to identify somatosensory deprivation as the primary etiological factor in the development of social-emotional disorders, particularly pathologic violent behaviors. Distinctions between near-receptor and distance-receptor functioning are drawn to emphasize their differential role in the psychobiological and psychosocial influences of early development. It is proposed that maternal-social deprivation experiences may be interpreted as a special case of partial functional somatosensory denervation and that CANNON's law of Denervation Supersensitivity may be the explanatory neurophysiological principal mediating sensory

deprivation effects. In particular, it is suggested
that the cerebellum becomes supersensitive and
hyper-excitable in function due to insufficient
somatosensory stimulation during early development.

464. Prescott, J.W. Sensory deprivation vs. sensory stimulation
during early development: A comment on Berkowitz's study. The
Journal of Psychology, 1971, 77, 189-191.

   Comments are offered to indicate that there does exist
   an extant theory of exclusive domain which can account
   for the findings of Berkowitz (Berkowitz, E.L. The effects
   of visual or auditory stimulation given in infancy
   upon subsequent preference for stimulation of both
   modalities. Get. Psychol. Monog., 1970, 81,
   175-196.) and, further, that the developmental sensory
   deprivation theory is quite specific with respect to
   expected effects of differing sensory system
   deprivation during early development. Far from
   being premature, this point of view has sufficiently
   matured to provide a useful explanatory system to
   account for the marked variations in the quality
   and quanity of early sensory experiences in understanding the developing brain and behavior.

465. Prescott, J.W. Body pleasure and the origins of violence.
The Futurist, 1975, April, 64-74.

   Aspects of this article are devoted to a discussion of
   Harlow's nonhuman primate research.

466. Prescott, J.W. Somatosensory deprivation and its relationship to the blind. In Z.S. Jastrzembska (Ed.), The effects of
blindness and other impairments on early development. New York:
The American Foundation for the Blind, 1976.

   The objective of this paper is to illustrate how
   a visual defect can result in deprivation to
   the somatosensory modality and that behavioral
   deficits usually attributable to the primary
   visual sensory deficit may, in fact, be
   attributable to deprivation of the somatosensory
   system. It is suggested that this type of

confounding may also occur when other sensory channels
suffer deprivation. Neurobiological and neurobehavioral
principles derived from animal sensory deprivation studies
are briefly summarized in order to clarify the meaning of
behavioral symptoms occurring in humans who also have been
deprived of sensory stimulation during early development.

467. Preston, D.G. Influences of the social and inanimate environment on the exploratory and play behavior of juvenile Java (M.
irus) monkeys. Dissertation Abstracts International, 1972, 32(7-
13), 4260, 48:548.

The effects of variations in the novelty of the inanimate
and social environments over 5-day periods were investigated. The Ss were 4 juvenile (22-24 mo.) monkeys. The Ss
were allowed to enter an initially unfamiliar cage either
singly or as a group. A second experiment involved reunit-
ing 3 male juvenile monkeys with their mothers for 5 days
following a 15-mo. separation. Results of the experiments
indicated that variations in the social environment were
much more effective than variations in the object environment
in changing the Ss's behavior. The presences of a strange
adult female monkey reduced activity significantly, while
the presence of the mother along with 2 strange females
resulted in little decrease. In several cases, social
behaviors were still increasing in frequency at the end
of each 5-day period, indicating that satiation proceeded
slowly for social activity. During the 5 days of reunion, a mother-offspring relationship was reestablished.
This relationship involved few overt interactions between
mother and son, in contrast to the amount of mother-infant
interaction in younger macaque monkeys following a period
of maternal separation. The relationship between behavioral
reactions to novelty and level of arousal were discussed.

468. Preston, D.G., Baker, R.P., & Seay, B. Mother-infant separation in the patas monkeys. Developmental Psychobiology, 1970, 3
(3), 298-306.

Studies of human beings and several macaque monkey species
have indicated a similar reaction to maternal separation,
involving a short-lived agitation, followed by a period of
physical and mental depression, and, in human beings,
sometimes culminating in social indifference and sociopathic behavior. It was found that the six 7-month-old

patas monkey infants used in this study showed the typical
initial agitation and subsequent decline in social behavior
during the 3-week separation; but that an increase in peer
interaction, rather than the common rise in maternal interaction found in other species, followed reunion. Three
factors may interact to produce a less severe separation
reaction and increased peer interaction following separation,
as compared to the results of macaque studies: (a) patas
tend to be more permissive mothers, (b) patas infants have
a longer, more intense, early mother-infant relationship,
and (c) contact play is a secondary play pattern in patas,
but is the primary play pattern in macaques.

469. Raisler, R.L. & Harlow, H.F. Learned behavior following
lesions of posterior association cortex in infant, immature, and
pre-adolescent monkeys. Journal of Comparative and Physiological
Psychology, 1965, 60(2), 167-174.

Bilateral lesions in temporal neocortex and pre- and
parastriate areas were produced in groups of monkeys
at 130, 370 and 900 days, respectively. Delayed response
performance of these groups was appropriate to age and
apparently normal. Consistent deviations from normal
patterned-strings performance and a deficit in object-
quality discrimination learning were observed only in
the 900-day group. The latter was not permanent and
recovery was attributed to the ability of these Ss to
relearn the use of color, but not form, cues in discrimination problems. In general, the results support the
hypothesis that sparing of function after ablations of
posterior association cortex is related to type of test
and age at the time of operation.

470. Randolph, M.C. & Mason, W.A. Effects of rearing conditions
on distress vocalizations in chimpanzees. Folia Primatologica, 1969,
10, 103-112.

Distress vocalizations of wild-born and socially-deprived
chimpanzees were compared in two experiments. In the
first experiment, the animals were observed in an
unfamiliar room while alone, and in the presence of a
human companion. In the second experiment, the animals
were given a shock for contacting a familiar object in
order to determine the effects of pain on vocal activity.

In both experiments, and under all conditions of testing,
the level of distress vocalizations was much higher in
wild-born animals. Two factors are considered in an
attempt to account for these results: The first emphasizes
the role of social feedback in strengthening the tendency
to respond vocally to stress; the second recognizes the
possibility that behaviors which develop in the socially-
deprived chimpanzee provide a way of coping with stress that
is not seen in the wild-born animal.

471. Rasch, E., Swift, H., Riesen, A.H., & Chow, K.L.   Altered
structure and composition of retinal cells in dark-reared mammals.
Experimental Cell Research, 1961, 25, 348-363.

This report describes some cytological and cytochemical
changes in retinas from cats reared under varying conditions
of light stimulation. Additional observations have been
noted for retinas from normal and dark-reared chimpanzees
and rats. These animals are part of a long-term project
to evaluate effects of early visual stimulation upon
later development of visually guided behavior. The data
suggest that adequate light stimulation is a major variable
controlling the development of normal ribonucleoprotein
levels in cells of the mammalian retina.

472. Redican, W.R., & Mitchell, G.A.   A longitudinal study of
paternal behavior in adult male rhesus monkeys, 1. Observations of
the first dyad. Developmental Psychology, 1973, 8, 135-136.

The present study was designed to assess affiliative
paternal behavior in rhesus monkeys. Adult male
monkeys were each paired with a one month old infant.
Preliminary results indicate strong attachments within
the paternal affectional system.

473. Redican, W.K. & Mitchell, G.   The social behavior of adult
male-infant pairs of rhesus monkeys in a laboratory environment.
American Journal of Physical Anthropology, 1973, 38, 523-526.

In the present study two pairs of adult male and infant
rhesus macaques were housed together without the mother
for a period of at least seven months. To minimize the
probability of physical harm to the infants, one adult
male was present at the birth of the infant with whom it
was paired and the other was visually familiarized with
the infant for several weeks before being paired with it.
Both pairs have shown much less ventral-ventral contact
than mother-infant pairs and the infants have rarely
attempted to gain nipple contact. The female infant now
initiates and maintains ventral-ventral contact much more
frequently than the male infant in spite of the higher
degree of aggression directed toward her by the male.
The adults frequently groom their infants and exhibit
varying degrees of both tolerance and aggression. Rough-
and tumble-play appeared quite early in both pairs and is
much more vigorous than mother-infant play. Recently
the male-male pair has exhibited reciprocal mounting
and thrusting. Comparisons are made with data obtained
in an earlier study of mother-infant pairs conducted
under nearly identical experimental circumstances. The
effects of separating the adult males and infants at
seven months are also described.

474. Redican, W.K., Gomber, J., & Mitchell, G. Adult male
parental behavior in feral- and isolation-reared rhesus monkeys
(Macaca mulatta). In J.H. Cullen (Ed.), Experimental behavior: A
basis for the study of mental disturbance. Dublin: Irish University Press, 1974, 131-146.

This paper emphasizes the development of an infant
male who was paired with a feral-born adult male, and
of an infant female who was paired with an isolation-
reared adult male. Comparisons are made with a completed
study of four mother-infant pairs under similar circumstances.
In all studies contact with peers was prohibited to rule out
any interactive effects of peer rearing. Throughout data
collection, a time-sampling technique was used to measure
frequency and duation of behaviors, and detailed written
records were also taken.

475. Regal, D.M., Boothe, R., Teller, D.Y., & Sackett, G.P. Visual
acuity and visual responsiveness in dark-reared monkeys (Macaca
nemestrina). Vision Research, 1976, 16, 523-530.

> Infant pigtail macaques were reared in darkness during
> the first 3 or 6 months after birth. Following dark
> rearing, the animals were relatively unresponsive
> on a series of informal visual tests (placing, startle,
> etc.). Testing with a forced-choice preferential
> looking task showed acuities between 2.5 and 7.5
> c/deg (20/80 Snellen) in all dark reared monkeys
> compared with 15 c/deg (20/40 Snellen) in control
> animals. These results suggest that in macaque
> infants dark rearing is more harmful to visual responsive-
> ness than to visual acuity per se.

476. Reite, M. Maternal separation in monkey infants: A model of
depression. In I. Hanin and E. Usdin (Eds.), Animal models in
psychiatry and neurology. Oxford & New York: Pergamon Press, 1977,
127-140.

> The purposes of this paper as outlined by the author
> includes four separate but related topics in the field
> of nonhuman primate behavioral development. First, the
> author presents a detailed examination of certain con-
> ceptual problems involved in using animal models. Here
> he discusses the limitations of generalizing nonhuman
> data to human topics and the different types of animal
> models which have been proposed for this purpose. Second,
> an outline of the overall systems approach the author
> and others have developed in their laboratories for studying
> physiology and behavior in unrestrained group living
> monkey infants is developed. A third area covered is
> a summary of the data obtained to date on the physiology
> of the agitation-depression reaction in infant pigtail
> monkeys (M. nemestrina). Here the author reviews several
> major studies developed in his own laboratory and those of
> others. Finally, brief discussion is given to the potential
> advantages of this type of study to the enhanced under-
> standing of the human condition and how this model of
> depression in nonhuman primates might relate to human
> affective disturbances. The implications of this research
> for the field of psychiatry, psychology and physiology
> are emphasized.

477. Reite,   M. , & Short, R.A.   Nocturnal sleep in separated monkey infants.   Archives of General Psychiatry, 1978, 35, 1247-1253.

    Nocturnal sleep was recorded from ten unrestrained,
    group-living Macaca nemestrina (pigtail) monkey infants,
    using implantable multichannel biotelemetry systems, during
    the agitation-depression behavioral reaction that
    follows maternal separation. Sleep disturbances
    during the four nights of separation were characterized
    by decreases in rapid eye movement (REM) time and in
    the number of REM periods, and increases in REM latency.
    Time awake and number of arousals were increased. Slow-
    wave sleep was not significantly affected. Sleep pattern
    changes were most pronounced the first separation
    night, and tended to decrease as separation continued,
    whereas behavioral measures of depression tended to
    increase as separation continued (up to four days).
    Sleep patterns returned to normal following reunion
    with the mother. Those infants who had the most
    severe sleep disturbances the first separation night
    (more time awake, less total sleep, less REM) also
    tended to become most depressed behaviorally later in
    the separation period.

478.   Reite, M. , Kaufman, I.C., Pauley, J.D., & Stynes, A.J.
Depression in infant monkeys: Physiological correlates. Psychosomatic Medicine, 1974, 36(4), 363-367.

    This report presents certain physiological data including
    heart rate (HR), body temperature (BT), and sleep
    patterns during 36 hours of maternal separation in
    four pigtail (M. nemestrina) infants. Before
    separation BT circadian rhythms and sleep stage measures
    tended to show little intersubject variability, while
    HR values were much more variable and individually
    characteristic of a given monkey infant. As expected,
    HR was positively correlated with activity level.
    Immediately following separation, the infants were
    behaviorally agitated, and this agitation was
    accompanied by pronounced elevations of both HR and
    BT. In addition, during the first night following
    separation, sleep was markedly disturbed with all
    infants showing both a decrease in REM sleep and
    profound hypothermia.

479.   Reite, M., Short, R. Kaufman, I.C., Stynes, A.J., & Pauley,
J.D. Heart rate and body temperature in separated monkey infants.
Biological Psychiatry, 1978, 13(1), 91-105.

    Heart rate (HR) and body temperature (BT) were
    recorded from ten unrestrained group-living
    M. Nemestrina (pigtail) monkey infants, using
    totally implantable multichannel biotelemetry
    systems, during a 4-day base line (preseparation)
    period, during the agitation-depression reaction
    accompanying 4 days of maternal separation, and
    for several days following reunion with the mother.
    Quantified behavioral data were collected in nine
    of the ten infants. Mean daytime (1000 to 1600 hr)
    and nightime (2200 to 0400 hr) HR and BT values
    were computed for each infant, and for the group
    as a whole. The behavioral agitation reaction
    immediately following separation was accompanied
    by increases in both HR and BT. Beginning with
    the first night of separation, both HR and BT showed
    marked decreases from baseline.. Whereas group mean
    HR and BT changes were maximal early in the separation,
    behavioral indices of depression tended to be maximal
    later in the separation period. Reunion with the
    mother tended to normalize HR and BT in most infants.
    Two infants exhibited sudden transient reversible
    drops in nocturnal BT well into the separation period,
    suggesting an impairment in thermoregulatory mechanisms
    during the period of depressive behavior.

480.   Riesen, A.H. Effects of stimulus deprivation on the
development and atrophy of the visual sensory system. American
Journal of Orthopsychiatry, 1959, 30, 23-36.

    Prolonged stimulus deprivation during infancy in the
    chimpanzee was found to result in atrophy of the
    ganglion cell layer of the retina and the optic
    nerve. Although reversible during postnatal months,
    these functionally produced changes in the neural
    substrate became irreversible in 3 animals receiving
    little or no light stimulation for 16 months or longer.
    The results of several investigations are discussed
    with regard to transneuronal changes in the central
    nervous system and were related to the concept of an
    interdependence between neural growth, neural
    metabolism, and afferent stimulation.

481. Riesen, A.H. Brain and behavior: Session I. Effects of
stimulus deprivation on the development and atrophy of the visual
sensory system. American Journal of Orthopsychiatry, 1960, 30,
23-36.

Prolonged stimulus deprivation during infancy
in the chimpanzee was found to result in atrophy of
the ganglion cell layer of the retina and the
optic nerve. Although reversible during the
early postnatal months, these functionally
produced changes in the neural substrate
became irreversible in 3 amimals receiving
little or no light stimulation for 16 months
or longer. The histologic and cytochemical
nature of such effects of deprivation were
studied in cats which were given no light, 1
hour of light per day, or normal light for
varying periods after birth up to 40 months.
Comparative data were obtained from 2
chimpanzees and 6 rats. The results were
discussed in connection with other findings
on transneuronal changes in the central nervous
system and were then related to the concept
of an interdependence between neural growth,
neural metabolism, and afferent stimulation.

482. Riesen, A.H. Stimulation as a requirement for growth and
function in behavioral development. In D.W. Fiske and S.R. Maddi
(Eds.) Functions of varied experiences. Homewood, Illinois:
Dorsey Press, 1961, 57-80.

This article emphasizes the sensory and perceptual
aspects of early stimulus deprivation. The growth
and development of neural structures is highly
dependent upon early stimulation. The capacity
to process and assimilate complex sensory input,
within the visual modality, is markedly inferior
when only simplified stimulation is permitted to
enter that system during infancy. Behavioral
and biochemical correlates of this procedure are
discussed with regard to the ontogenesis of
various animal species, including primates.

483. Riesen, A.H. Studying perceptual development using the
technique of sensory deprivation. Journal of Nervous and Mental
Diseases, 1961, 132, 21-25.

  This article presents the results of two experiments,
  one done with a monkey which was deprived of visual
  stimulation from the age of five months to the age
  of 10 months, the other on a chimpanzee deprived of
  pattern light from 10 to 18 months of age. Both
  animals showed distinct deficits in visual perception,
  as well as eye-hand coordination.

484. Riesen, A.H. Effects of early deprivation of photic stimulation. In S.F. Osler and R.E. Cooke (Eds.), The biosocial
basis of mental retardation. Baltimore: Johns Hopkins Press,
1965, 61-85.

  The effects of stimulus deprivation on the neural
  development of rhesus monkeys and chimpanzees are
  discussed. The concepts of cross-modality
  transfer and discrimination learning are analyzed
  via this methodology.

485. Riesen, A.H. Sensory deprivation. In E. Stellar and J.M.
Sprague (Eds.), Progress in physiological psychology (Vol. 1).
New York: Academic Press, 1966, 117-146.

  The present paper discusses the relationship between
  functional demands and the development and maintenance
  of neural systems. Specifically, the effects of
  sensory deprivation on the micro-structure of the
  visual system are discussed in various animal species.

486. Riesen, A.H. Nissen's observations on the development of
sexual behavior in captive-born, nursery-reared chimpanzees. The
Chimpanzee, 1971, 4, 1-18.

  Henry W. Nissen, on the basis of many years of
  observations, proposed the following hypothesis:
  On a background of the necessary neuro-
  endocrine sexual maturation a social learning
  process is critical for the full development of
  chimpanzee reproductive behavior. As a test of

this hypothesis he kept nine male and seven female
chimpanzees under observation during all mating
opportunities for hundreds of hours over a two-
year period. No complete mating pattern occurred
in any pairings of previously inexperienced animals.
The failures and long delayed successful trial-and-
error learning were associated with early individual
living in a nursery condition. The period from 3
months to 2½ years is implicated as essential for
the development of basic social-emotional skills.
Other evidence also supports the conclusion that
mother-rearing and peer group social interactions,
as opposed to social restriction, during chimpanzee
childhood are significant in prompt and perceptive
exploration, effective social learning, and normal
speed of problem solving. The data have wider
implications for research in comparative behavioral
development.

487. Riesen, A.H., Dickerson, G.P., & Struble, R.G. Somatosensory
restriction and behavioral development in stumptail monkeys.
Annals of the New York Academy of Sciences, 1977, 290, 285-294.

The present experiments on 16 stumptail macaque
(M. speciosa) monkey infants are an attempt to
relate bizarre behaviors resulting from somatosensory deprivation to early brain changes. The
authors specifically wanted to determine whether
moderate rather than extreme degrees of somatosensory deprivation could be related to changes
in relevant sensory and motor systems of the
brain. Preliminary results indicate changes
in the number of dendritic branchings in the
prefrontal lobes and in motor cortex I and II
for those animals subjected to deprivation.
To what extent there is permanent brain "damage"
remains to be determined. These results point
to the need for a more thoroughgoing search for
the full extent and nature of changes that
undoubtedly go beyond those thus far identified.

488. Riesen, A.H., Ramsey, R.L., & Wilson, P.D. Development of
visual acuity in rhesus monkeys deprived of patterned light during
early infancy. Psychonomic Science, 1964, 1(2), 33-34.

Six monkeys given diffused light from birth to 20
or from birth to 60 days of age were found to have
low visual acuity when first tested at these ages.
With experience in patterned light their acuities
improved consistently and at a rate comparable to
that of normally reared monkeys.

489. Rogers, C.M., Davenport, R.K. Sexual behavior of differen-
tially reared chimpanzees. Proceedings of the 2nd International
Congress of Primatology, Atlanta, GA., 1968, 1, 173-177. (Karger,
Basel, New York, 1969)

The social behavior of chimpanzees reared under
highly restricted conditions for the first three
years of life was observed for 13 weeks under a
variety of social situations. Observations were
taken of the animals' sexual behairo and reactions
to sexual advances made by conspecifics of the
opposite sex. The results demonstrated that
distortions of the environment during infancy
and early childhood lead to decrements in the
sexual behavior of adult chimpanzees. Total
isolation appears to be less detrimental to sexual
behavior than substitute human maternal care.
Plasticity to this behavior is indicated in that some
isolation-reared chimps can learn to copulate in contrast
to reports on the rhesus monkey.

490. Rogers, C.M., & Davenport, R.K. Effects of restricted rearing on sexual behavior of chimpanzees. Developmental Psychology,
1969, 1(3), 200-204.

Five male and seven female chimpanzees were reared in
isolation for the first 3 years of life. During and
after puberty these chimpanzees were observed in pair
and group social stituations, sometimes with each other
and sometimes with wild-born animals. In contrast to the
isolation-reared rhesus monkey, most of these chimpanzees
learned to copulate with sophisticated partners. When
the performance of these chimpanzees is compared

with that of a group of human-reared chimpanzees, it
appears that the substitution of human interaction
during infancy is more detrimental to sexual
development than is the absence of any interaction.

491. Rogers, C.M. & Davenport, R.K. Intellectual performance of
differentially reared chimpanzees: III. Oddity. American
Journal of Mental Deficiency, 1971, 75, 526-530.

Six champanzees separated from their mother at
birth and raised in restricted environments
during infancy were compared to 7 Wildborn
enriched-environment chimpanzees on an
oddity learning task administered when
both groups were between 10 and 12 years
of age. Restricted subjects were inferior
throughout, but showed evidence of improvement.
Differences between the groups are accounted
for in terms of relative differences in
adaptability.

492. Rose, R.H., Mason, J.W., & Brady, J.V. Adrenal response to
maternal separation and chair adaptation in experimentally-
raised rhesus monkeys. Proceedings of the 2nd International Congress
of Primatology, 1968, 1, 211-218.

Infant male rhesus monkeys were raised in two
experimental conditions from birth to
three years of age. Four animals were housed
as an isolated mother-infant pair and seven
were raised in total tactile and visual
separation from other animals. At three years
the mother-raised animals were separated from
mother and later both groups were placed in
primate restraining chairs. Urine was collected
and analyzed for adrenal corticosteroids
(17-OHCS) and catecholamine excretion. In
both groups animals showed significant differences
between one another in response to both separation
and placement in the restraining chair. Among
the mother-raised animals the differences in
endocrine response were maintained across both
stress situations, one animal had consistently
the greatest response while another showed no
response in either situation. No significant
differences in psychoendocrine responses were

observed when group means were compared. In conducting
the research reported herein, the investigators
adhered to the 'Guide for Laboratory Animal Facilities
and Care' prepared by the Committee on the Guid for
Animal Resources, National Academy of Science, National
Research Council.

493.   Rosenblum, L.A.   The development of social behavior in the
rhesus monkey, (Doctoral dissertation, University of Wisconsin,
1961).   Dissertation Abstracts International, 1961, 22(3).

Eight rhesus monkeys (M. mulatta) were separated from
their mothers 6-12 hours after birth and raised in
individual cages with artificial mother surrogates, and
their behavior observed for the first 6 months of
life. Social behavior development observed in this
study was similar to that described by previous authors
for species both above and below the rhesus monkey on the
phyletic scale. The fact that the behavior of the subjects
was quite similar to behaviors observed both in the
wild and in other laboratory situations implies that the
basic forms of development which characterized these subjects
probably hold considerable generality to the species as
a whole.

494.   Rosenblum, L.A.   Mother-infant relations and early behavioral
development in the squirrel monkey.   In L.A. Rosenblum and R.W.
Cooper (Eds.), The Squirrel Monkey.   New York: Academic Press,
1968.

This chapter considers the ontogeny of the squirrel
monkey by reviewing pertinent published literature and
focusing in quantitative detail on a description of the
first 10 months of development of several laboratory-
born squirrel monkey infants of the Peruvian type. Both
mother-infant relations and infant behavioral development
is discussed, and an attempt is made to compare the
squirrel monkey's development with that of the infants
of bonnet and pigtail macaques (M. radiata and M. nemestrina),
which have been studied under similar conditions.

495. Rosenblum, L.A. Infant attachment in monkeys. In R. Schaffer (Ed.), <u>The origins of human social relations</u>. New York: Academic Press, 1971, 85-113.

The current paper presents a series of data drawn from a number of related comparative studies on infant attachment. Each study bears upon several hypotheses derived from the child literature regarding factors which may influence the level and endurance of infant-mother attachment patterns. These studies on a number of nonhuman primate forms lend support to the hypothesis that phenomena relating to human infant attachment rest upon a meaningful biological base. Moreover, the pertinent evolutionary antecedents of childhood attachment are evident and available for study in nonhuman primates in ways often difficult or ethically impossible to carry out at the human level.

496. Rosenblum, L.A. The ontogeny of mother-infant relations in macaques. In H.S. Moltz (Ed.) <u>Ontogeny of vertebrate behavior</u>. London: Academic Press, 1971, 315-367.

The ontogeny of the macaque genus is discussed from the mother-infant affectional system. Maternal experience with regard to parturition as well as, mother-infant interactions such as contact, maternal separation, nursing, protective behavior, and maternal rejection are discussed.

497. Rosenblum, L.A. Maternal regulation of infant social interactions. In C.R. Carpenter (Ed.), <u>Social regulatory mechanisms of primates</u>. University of Pennsylvania Press, 1973, 195-217.

This paper, through the use of laboratory observations, has several objectives: to describe the ways in which mothers in several species--pigtails (<u>Macaca nemestrina</u>), bonnets (<u>Macaca radiata</u>) and squirrel monkeys (<u>Saimiri sciureus</u>)--regulate the amount and character of social interaction of their infants with other group members; to describe some of the factors which influence the shaping of variations in maternal regulation; and to suggest some of the long-term sequelae of the consequent differences in early social experiences of the infants.

498. Rosenblum, L.A. Sex differences, environmental complexity
and mother-infant relations. Archives of Sexual Behavior, 1974,
3(2), 117-128.

Differential handling of male and female infants
by their mothers and differential behavior of
male and female infants independent of maternal
behavior is a critical research issue with significant
implications for understanding human development.
This report describes mother-infant interaction in
the squirrel monkey as a function of the infant's
sex in a variety of laboratory environments. In
a complex social environment, males move away
from their mothers at a younger age than do
females. The mothers then attempt to keep the
males physically close. This maternal effort
at continuing mother-son closeness is not successful,
indicating greater infant-activated autonomy in the
male.

499. Rosenblum, L.A. Sex differences in mother-infant attachment
in monkeys. In R.L. Vandewiele, R.M. Richart and R.C. Fried
(Eds.), Sex differences in behavior. New York: Wiley, 1974,
123-145.

This chapter reports data from two studies concerned
with sex differences in early mother-infant relations
in monkeys. Results indicate that shortly after the
age at which infants begin excursions from the
mother and experience more overt contact with the
outside environment, male infants show a greater
readiness to move toward moderately novel, complex, or
arousing stimuli than do females.

500. Rosenblum, L.A., & Alpert, S. Fear of strangers and
specificity of attachment in monkeys. In M. Lewis and L.A.
Rosenblum (Eds.) The origins of behavior: The origins of fear.
New York: Wiley, 1974.

This chapter gives an overview of the research
on mother-infant attachment and fear of strangers

in nonhuman primates. Research by the authors is
described which was initiated in light of the
pivotal significance for primate development
inherent in the establishment of selective
responsiveness to the mother and related or
opposed reactions to others. These studies
focused on two macaque species, the bonnet
(Macaca radiata) and the pigtail (Macaca nemestrina),
and data is presented on mother-infant relations
and infant development for the two species.

501. Rosenblum, L.A., & Harlow, H.F. Approach-avoidance conflict
in the mother-surrogate situation. Psychological Reports, 1963,
12, 83-85.

During the first five and one-half months of life
two rhesus monkeys were intermittently blasted
with compressed air, an aversive stimulus, while
contacting a cloth surrogate. They spent significantly
more time on the surrogates than four control infants
having equal access to the standard cloth surrogate.
The results contrast with expectations based on Neal
Miller's formulation of approach-withdrawal
conflict behavior. The generality of Miller's
formulation is therefore questioned.

502. Rosenblum, L.A., & Harlow, H.F. Generalization of
affectional responses in rhesus monkeys. Perceptual and Motor
Skills, 1963, 16, 561-564.

The present study demonstrated the ability of 8 young
rhesus monkeys to generalize affectional responses to
a surrogate stimulus which differed in size from the
original and failed to provide contact-comfort.

503. Rosenblum, L.A., & Kaufman, I.C. Laboratory observations
of early mother-infant relations in pigtail and bonnet macques.
In S.A. Altman (Ed.), Social communication among primates.
Chicago: University of Chicago Press, 1967, 33-41.

Considering the widespread interest in social
behavior and its development throughout
the primate order as evidenced by the diversity

of work included in this book, we can see the need
for close comparisons between primate species studied
under comparable and relatively controlled conditions.
As part of a laboratory program designed to assess the
role of varying patterns of mother-infant relations
upon the behavioral development of the young, during
the past 4 years we have been studying homospecific
groups of two species of macaque, the pigtail (Macaca
nemestrina) and the bonnet (M. radiata) under
identical social and environmental conditions.

504. Rosenblum, L.A., & Kaufman, I.C. Variations in infant
development and response to maternal loss in monkeys. American
Journal of Orthopsychiatry, 1968, 38(3), 418-426.

Sustained physical contact between animals
characterizes bonnet but not pigtail monkey
groups. This pattern encourages maternal
permissiveness and enhances social orientation
in infants. Moreover, pigtail infants show
depression and behavioral debilitation after
maternal loss, whereas bonnets are often
adopted by others and suffer minimal affective
and behavioral changes.

505. Rosenblum, L.A., & Lowe, A.C. The influence of familiarity
during rearing on subsequent partner preferences in squirrel
monkeys. Psychonomic Science, 1971, 23, 35-37.

Laboratory-born adolescent squirrel monkeys,
drawn from four separate rearing groups, were
placed together. Strong consistent preferences
for interacting with others of the same rearing
group were observed in 6 weeks of initial
observation and persisted in follow-up
observations 14 months later. Play was the
major type of interaction between strangers and
may serve as the primary mechanism through which
social integration occurs.

506.   Rosenblum, L.A., & Youngstein, K.P.   Developmental changes in
compensatory dyadic response in mother and infant monkeys.   In
M. Lewis and L.A. Rosenblum (Eds.), The origins of behavior--
The effect of the infant on its caregiver.   New York: John Wiley
and Sons, 1974, 141-161.

     This chapter describes a study which attempts to
     delineate further the differential role of mother
     and infant in the developing dyadic relationship
     in nonhuman primates.   The results suggest that
     the contribution made by infant bonnet macaques
     (M. radiata) and mothers descreases normally over
     time, but more rapidly in mother than in infant;
     the most rapid decreases occur in the first 3 to
     6 months of life, approaching a gradually
     declining plateau at about 4 to 6 months in the
     mother and 8 to 10 months in the infant.   Further,
     each partner's capacity to compensate for the
     lack of appropriate contribution from the other
     decreases in a linear fashion after about the
     second to third month of the infant's life.
     The model suggests that by about the end of
     the first year of life each partner loses the
     capacity to compensate for a failure on the
     part of the partner to respond.

507.   Rosenblum, L.A., Clark, R.W., & Kaufman, I.C.   Diurnal
variations in mother-infant separation in two species of macaque.
Journal of Comparative and Physiological Psychology, 1964, 58(2),
330-332.

     The degree of cohesiveness within mother-infant
     dyads of Macaca nemestrina (pigtails) and
     Macaca radiata (bonnets) measured in terms of
     the physical separation between them, was studied
     in laboratory groups throughout 24-hr. periods.
     In both species cohesiveness was greatest at
     5 A.M. when sleeping was at its peak.   The pigtail
     dyads were significantly less cohesive prior to
     feeding.   They also showed a greater tendency to
     assume a recumbent posture during sleep.   These
     differences in dyadic patterning appear consistent
     with observed differences in the general behavior
     of these species.

508.   Rosenblum, L.A., Coe, C.L., & Bromley, L.J.   Peer relations
in monkeys: The influence of social structure, gender, and familiarity.   In   M. Lewis and L.A. Rosenblum (Eds.), The origins of
behavior: Peer relations and friendships.   New York: Wiley,
1975,   67-98.

This chapter examines a number of the most salient
features of peer relations and the foundations of
these companionships during early and later infancy
and adolescence in a number of nonhuman primate
species. Comparative laboratory studies by the
authors on peer relations are discussed with
particular focus on the role of species
differences and the likely impact of species-
characteristic social structures, the role of
gender in shaping the form and selectivity of
peer relations, the influence of familiarity
between peers, and the complex interactions
of these factors in molding the course of peer
relations. By providing a glimpse of the
evolutionary continuity of such patterns across
several primate species, the authors hope to
suggest the general significance of congruent
features of peer relations within the primate
order.

509.   Rowell, T.E.   The effect of temporary separation from their
group on the mother-infant relationship of baboons.   Folia
Primatologica, 1968, 9, 114-122.

Interactions between mother and infant baboon
were observed during their second month in
three social environments: in the group
in which they normally lived, separated from
the group ('isolated') and confined with one
other group member at a time ('simplified').
Mothers and infants were only in the last
two situations for short periods, but showed
a change in behaviour relative to their behaviour
in the group, similar to that reported in long-
term isolation; they restricted their babies less,
babies played more on or near their mothers, and
mothers groomed their infants more. All changes
except that in mothers grooming infant which
were seen in 'isolated' were also seen in the
'simplified' test situation. This finding is
related to the difference between social

environments in a cage and in the wild.

510. Rowell, T.E., & Hinde, R.A. Vocal communication by the
rhesus monkey (Macaca mulatta). Proceedings of the Zoological
Society of London, 1962, 138, 279-294.

Noises are difficult to describe, and this has
been a barrier to understanding their meanings
and using them in studies of behavior. In the
present study tape recordings and spectrographs
were used to learn about the properties of noises
made by rhesus macaques. Noises fall into two
groups, harsh noises and clear calls, which are
considered separately. Eight harsh noises associated
with agonistic behavior and four occurring in friendly
situations are described. Clear calls show more
individual variation and occur during various
situations. A wide range of intermediaries between
the basic noises make a large "vocabulary" available
to express small differences in mood.

511. Rowell, T.E., & Hinde, R.A. Responses of rhesus monkeys to
mildly stressful situations. Animal Behavior, 1963, 11, 235-243.

Rhesus monkeys were confronted with a series of
situations intended to provide a controlled version
of mildly stressful occurrences in everyday life.
Detailed records of their behaviour in these situations
(offered food, and watched by a familiar person and a
person in a grotesque mask) were compared with their
behaviour when undisturbed. An operational definition
of the stressful situation could be made, for this context,
in terms of the increases in some types of behaviour and
decreases in others. All animals were tested in their
normal social environment and after six hours isolation.
The main effect of isolation was to intensify the stressful nature of the test situations. Differences between
age and sex classes are discussed briefly.

512. Rowland, G.L. The effects of total social isolation upon
learning and social behavior in rhesus monkeys. (Doctoral
dissertation, University of Wisconsin, 1964). Dissertation
Abstracts International, 1965, 64-6960, 1364-1365.

Rhesus monkeys were reared in total social isolation.
Three conditions of isolation were employed: birth
until 6 mos., 6 mos. until one year, and birth until
one year. These animals were extensively tested to
determine the effects of total social isolation upon
intellectual and social behaviors. The tests utilized
were: home cage observations, formation of object quality
discrimination learning set, 0 and 5 sec. delayed
response, avoidance conditioning, social testing in
playroom situations, and dominance hierarchy determinations.
While in isolation, those Ss isolated from birth were
characterized by extreme disturbance in the presence of
novel stimuli. These patterns of disturbed behaviors
were idiosyncratic and usually involved stereotyped
movements and immobility. These Ss also failed to establish
normal diurnal cycles until exit from isolation. There
were no differences between controls and isolates on
any of the learning tasks. The conclusions drawn was
that isolation has no measurable effect upon intellectual
processes. On the other hand, monkeys isolated from
birth until 6 mos. displayed deficits in social behaviors;
monkeys isolated from birth until 1 yr. showed almost
no social responsiveness. Monkeys isolated from 6 mos.
until 1 yr. were similar to their control animals. The
conclusion was drawn that social experience gained within
the first year of life is necessary for the development
of normal social responses in the monkey.

513. Roy, A.M. Early rearing of infrahuman primates with
reduced conspecific contacts: A selected bibliography. Part I.
Journal of Biological Psychology, 1976, 18(2), 36-42.

The first part of a comprehensive list of articles
(122 of 248) which deal with atypical rearing experiences
in infrahuman primates, especially rhesus monkeys and
chimpanzees, is presented. Articles are included which
present or evaluate the behavioral and physiochemical
consequences of rearing infants with varying degrees of
reduced contact with their species members early in
life.

514. Roy. A.M. Early rearing of infrahuman primates with reduced
conspecific contacts: A selected bibliography. Part II. Journal
of Biological Psychology, 1977, 19(1), 21-27.

The second part of a comprehensive list of articles
which deal with atypical rearing experiences in
infrahuman primates, especially rhesus monkeys and
chimpanzees, is presented. Articles are included
which present or evaluate the behavioral and
physiochemical consequences of rearing infants with
varying degrees of reduced contact with their species
members early in life.

515. Roy. A.M. Early rearing of infrahuman primates with
reduced conspecfic contacts: A selected bibliography.   Part III.
Journal of Biological Psychology, 1977, 19(2), 29-30.

The third part of a comprehensive list of articles
which deal with atypical rearing experiences in
infrahuman primates, especially rhesus monkeys and
chimpanzees, is presented. Articles are included which
present or evaluate the behavioral and physiochemical
consequences of rearing infants with varying degrees
of reduced contact with their species members early
in life.

516.   Rumbaugh, D.M. Maternal care in relation to infant behavior
in the squirrel monkey. Psychological Reports, 1965, 16, 171-176.

Observation of a mother squirrel monkey with her dead
infant and an experiment with another mother and her
viable 10-day-old infant are used to support the
following points. (1) The typical, passive maternal care
given by the squirrel monkey mother to her young is in
part influenced by the behavior of the infant. (2) In
response to certain behaviors of the infant, the
squirrel monkey mother can exhibit behaviors that for
her are otherwise improbable, including bipedal carrying
and cradling of the infant. Implications of these
findings for inferring the origin of bipedalism in primates
are considered.

517.   Rupenthal, G.C., Arling, G.L., Harlow, H.F., Sackett, G.P.,
& Suomi, S.J. A 10-year perspective of motherless-mother monkey
behavior. Journal of Abnormal Psychology, 1976, 85(4), 341-349.

Previous studies have reported that the maternal behavior

of rhesus monkey females who themselves were reared
without mothers ("motherless mothers") is generally
inadequate and often abusive. The present study
examined the maternal competency of 50 such subjects
with respect to the variables of rearing environment,
age at first social contact, sex of offspring, age at
first delivery, parity, and duration of exposure to
previous offspring. The results suggested that physical
contact with conspecifics, either with peers prior to
adulthood or with their own infants immediately
after birth, greatly reduced the probability that
motherless mothers would be inadequate maternally.

518. Ruppenthal, G.L., Harlow, M.K., Eisele, C.D., Harlow, H.F.,
& Suomi, S.J. Development of peer interactions of monkeys reared
in a nuclear family environment. Child Development, 1974, 45,
670-682.

Rhesus monkeys were reared for the first 3 yr. of life
in a nuclear family environment which permitted continual
access to their mothers and fathers and limited access
to peers and to other adults of both sexes. The subjects
rapidly developed sophisticated patterns of social
behavior seldom observed in laboratory-reared monkeys
and maintained levels of interactive play longer
chronologically than has been reported for feral-
raised animals. Sex differences with respect to
grooming and play behaviors emerged early in life
and were maintained by the present subjects. Implications
of the findings are discussed.

519. Sackett, G.P. Effects of rearing conditions upon the
behavior of rhesus monkeys (Macaca mulatta). Child Development,
1965, 36(4), 855-868.

Experiments are reviewed indicating that stimulus deprivation during rearing can produce monkeys that are
inactive, prefer visual and manipulatory stimuli of low
complexity, show little exploration of the environment,
are sexually and maternally abnormal, and generally
withdraw from social contact. These behaviors are
explained by a complexity dissonance preference theory,
which assumes normal behavioral development proceeds by
a gradual process of paced increments in environmental
complexity. Aspects of abnormal behavior in monkeys are
explained in terms of inappropriate pacing of

environmental complexity. Data supporting this theory,
as well as data revealing long-lasting effects of
stimulation during the first month of life, are
presented.

520. Sackett, G.P. Manipulatory behavior in monkeys reared under
different levels of early stimulus variation. Perceptual and Motor
Skills, 1965, 20(3), 985-988.

Preference for proprioceptive stimulation was tested
in monkeys reared in (a) the jungle, (b) a wire cage,
and (c) an enclosed metal cage. The first group had
the greatest level of stimulus input early in life, the
last group had the lowest level. The frequency and
duration of contact during 12-hr. trials, with a chain
a movable T-bar, and a nonmovable bar were measured. Ss
were at least 3 yr. old when tested. Animals raised in
the jungle had the highest frequency and duration of
contact with all three stimuli. However, as the
proprioceptive complexity inherent in the stimulus
decreased, manipulation decreased for jungle-reared
animals but increased for laboratory-reared monkeys.
It was concluded that early deprivation of tactual-
proprioceptive stimulation produces an animal who
prefers manipulatory stimuli of low proprioceptive
complexity later in life. This interpretation was
based on a stimulus complexity preference hypothesis
that was used to explain the failure of many laboratory-
born monkeys to learn simple manipulatory responses in
instrumental learning situations.

521. Sackett, G.P. Response of rhesus monkeys to social stimulation presented by means of colored slides. Perceptual and Motor
Skills, 1965, 20(3), 1027-1028.

This report presents preliminary data on the effects
of colored slides as social stimuli for rhesus monkeys
(Macaca mulatta). The results suggest two conclusions.
(1) Colored slides with different contents produce
different responses in rhesus monkeys. (2) The
level of response to a given slide can vary with age
and/or rearing condition. Thus, response to a
photographic stimuli may be used to assess effects
of differential early rearing, as well as probe the
particular locus of response deficits produced by
developmental and other variables.

522. Sackett, G.P. Development of preference for differentially
complex patterns by infant monkeys. Psychonomic Science, 1966, 6
(9), 441-442.

> Rhesus monkeys, reared in isolation were exposed to
> afferentially complex visual patterns from Day 5
> through 39 after birth. Duration of visual and manual
> exploration served as a measure of stimulus perference.
> As the infant monkeys matured a gradual change in
> preference from simple to more complex stimuli was
> observed. Compared with human infant behavior these
> data extend the phylogenetic generality of stimulus
> complexity as an important motivational variable in
> behavioral development, and reinforce the assumption
> that stimulus preferences proceed from simpler to more
> complex input as a function of age.

523. Sackett, G.P. Monkeys reared in isolation with pictures as
visual input: Evidence for an innate releasing mechanism. Science,
1966, 154, 1468-1473.

> Monkeys reared in isolation from birth to 9 months
> received varied visual input solely from colored slides
> of monkeys in various activities and from nonmonkey
> pictures. Exploration, play, vocalization, and disturbance
> occurred most frequently with pictures of monkeys
> threatening and pictures of infants. From 2.5 to 4
> months threat pictures yielded a high frequency of dis-
> turbance. Lever-touching to turn threat pictures on was
> very low during this period. Pictures of infants and
> of threat thus appear to have prepotent general
> activating properties, while pictures of threat appear to
> release a developmentally determined, inborn fear response.

524. Sackett, G.P. Some effects of social and sensory deprivation
during rearing on behavioral development of monkeys. Revista
Interamericana de Psicologia, 1967, 1, 55-80.

> Rhesus monkeys (Macaca mulatta) were reared in (i) total
> social isolation with unchanging nonsocial stimulation
> for the first 3, 6, or 12 months of life, (ii) partial
> isolation, seeing and hearing but not touching, other
> monkeys, (iii) with peers but no mothers, and (iv) with
> peers and mothers. Tests throughout development indicate
> that early total isolation for 6 or 12 months produces
> lasting deficits in socialization and in nonsocial

behaviors. One year of isolation is particularly
devastating. The isolate monkey's learning ability,
however, is not retarded compared with other laboratory
born animals. Partial isolation also produces behavioral
deficits, but not to the extreme found in total isolation.
A sufficient condition for normal social development seems
to be peer-rearing, regardless of presence or absence of
maternal experiences. Rearing with several different peers
seems particularly potent as a socializing influence.
Data are also described for animals reared in total social
isolation, but receiving visual stimulation consisting of
pictures of monkeys. Results show that some forms of
visual social communication may be innately recognized, and
appropriately responded to, by infants who never contact
a real monkey. Pictures of monkeys threatening
produced avoidance behavior between 60 and 90 days, but
threat stimuli did not produce avoidance before this age.
This "innate releasing mechanism" thus appears to depend
on postnatal maturation before becoming operative. Pictures
of threat and pictures of infants generally appeared to be
prepotent for releasing high levels of many behaviors.
The postisolation behavior of visually stimulated isolates
seems more socially oriented and is clearly more
positively oriented toward the inanimate environment, than
the behavior of monkeys reared in isolation without varied
visual input. Although visual input appears to alleviate
some of the drastic effects of social isolation, at 1.5
years old visually enriched monkeys are grossly abnormal
compared to peer-raised animals.

525. Sackett, G.P. Some persistent effects of different rearing
conditions on preadult social behavior of monkeys. Journal of
Comparative and Physiological Psychology, 1967, 64(2), 363-365.

     Rhesus monkeys were reared (a) in total isolation for
     1 yr., (b) in total isolation for 6 mo., (c) in a
     bare-wire cage for 1 yr., (d) with normal mothers and
     peer experience for 9 mo., and (e) with maternally
     inadequate, brutal, motherless mothers and peer
     experience for 9 mo. At 4 yr., when groups were tested
     for social behavior with a male and a female stimulus
     monkey, differences in gross motor activity, aggression,
     fear withdrawal, and initiation of physical contact
     were found. In general, a rank-order relationship was
     obtained between amount of input early in life and pre-
     adult behavioral development. Experience with a hostile
     mother during the first months of life seemed to produce
     hyperaggression in 4-yr.-olds.

526. Sackett, G.P. Abnormal behavior in laboratory-reared rhesus
monkeys. In M.W. Fox (Ed.), Abnormal behavior in animals. Phila-
delphia: W.B. Saunders Co., 1968.

     Research in behavioral development leaves little doubt that
     adult behavior is influenced by early life experiences
     for animals. In species with simpler behavior, manipu-
     lations early in life seem to produce transient effects that
     do not persist into adulthood. In monkeys, however,
     differential early experiences have produced permanent
     abnormalities which span the total behavioral repertoire.
     This paper will review studies of behavioral development
     in rhesus monkeys (Macaca mulatta) conducted at the
     Wisconsin Primate Laboratory and will include a summary
     of the range of rearing conditions thus far studied, the
     types of behavioral situations and measurement
     techniques employed, and some of the behavioral abnormal-
     ities that have been observed.

527. Sackett, G.P. The persistence of abnormal behavior in
monkeys following isolation rearing. In R. Porter (Ed.), CIBA
Foundation symposium on the role of learning in psychotherapy.
London: Churchill, 1968.

     We have seen that early rearing of rhesus monkeys
     (Macaca mulatta) without physical peer contacts
     produces persistent and damaging abnormalities in
     nonsocial, social, sexual and maternal behaviours.
     These abnormalities appear to be uncorrelated with
     intellectual deficits. Attempts to reverse these
     effects of early experience have been uniformly
     negative. After extensive post-rearing social
     experience, animals deprived of physical peer contact
     during the first 3 to 6 months of life failed to
     perform socially at levels comparable to those
     achieved by animals reared with peer experience.
     Monkeys receiving visual social experience from pictures,
     but otherwise isolated from social contact, showed
     appropriate social behavior during rearing, but
     failed to develop normally after removal from isolation.
     Even when these monkeys were gradually paced in the
     introduction of novel and complex social and nonsocial
     stimulation, their behaviour was totally inadequate and
     failed to improve after repeated social experience.
     Thus isolation from critical early social and nonsocial
     experiences appears to produce permanent anomalies
     within the animal which persist regardless of the degree
     of adaptation to post-rearing test situations. Evidence

has also been presented to show that early learning,
perhaps within the first month of life, can produce
social traits, such as hyperaggression and preferences
for specific social stimuli, that persist into
adulthood. Finally, although an attempt to condition
increased physical contact in abnormal animals failed
to transfer to a social situation that did not
contain stimulus elements from the conditioning
situation, some evidence for the value of such
"behaviour therapy" approach was obtained. This was
particularly true for juvenile animals.

528. Sackett, G.P. Innate mechanisms, rearing conditions, and a
theory of early experience effects in primates. In M.R. Jones
(Ed.), Effects of early experience. Miami Symposium on the prediction of behavior, 1968. Coral Gable, Florida: University of
Miami Press, 1970, 11-53.

The purpose of this paper is to review selected studies
dealing with developmental factors influencing rhesus
monkey (Macaca mulatta) entogeny. The experimental
data deal with innate information processing functions in
primates, and the persistence into adulthood of effects
produced by variations in social and nonsocial
stimulation. Substantial emphasis is also placed on
individual differences related to gentic, prenatal,
perinatal, neonatal, infantile, and preadult factors.

529. Sackett, G.P. Unlearned responses, differential rearing
experiences, and the development of social attachments by rhesus
monkeys. In L.A. Rosenblum (Ed.), Primate behavior: Developments
in field and laboratory research (Vol. 1). New York: Academic
Press, 1970, 111-140.

The work reviewed in this paper does not generally
reinforce learning theories of social attachment aquisition,
at least not for monkeys. Although it does seem clear
that social attachment maintenance depends on feedback
during interactions with other animals, it appears that
this feedback must occur early in the animals life to be
effective. In these experiments social approach measures
used to index social preference yielded the following
information. (1) Neonate and infant monkeys separated from
their mothers at birth and having no access to animals
other than age-mates preferred adults to their own rather

than other species, preferred adult females over males,
and showed maturational changes in sex preferences that
were not related in any obvious manner to positive
reinforcement opportunities. (2) Social stimuli
available during the first days and months of infancy
were preferred by preadult animals 3-4 years after the
specific early experiences. (3) Like-reared monkeys
were preferred over animals reared under different
conditions even when the like-reared social stimuli were
completely unfamiliar and even though some rearing
groups had not experienced positive social interactions
with other animals. (4) Infants receiving adequate
maternal care preferred their own over other mothers,
but infants receiving inadequate maternal care involving
negative reinforcement showed greater preference for
their own mother than infants receiving species-typical
mothering. (5) Within 2 weeks of separation from the
mother, infants that were adequately mothered preferred
age-mates over adult females even though they had almost
no previous experience with age-mates. Inadequately
mothered monkeys did not respond to age-mates but
preferred adult females. (6) Adult female monkeys
preferred neonates over animals of other ages only
if they had been reared by a real mother. Thus these
data strongly suggest that inborn and maturational
perceptual and response mechanisms support the
acquisition of social approach behaviors in the absence
of specific conditioning and reward opportunities. These
results, however, also indicate that the maintenance of
social approach motivation depends on the quality of
feedback from social stimuli experienced during the
monkey's infancy.

530. Sackett, G.P. Exploratory behavior of rhesus monkeys as a
function of rearing experience and sex. Developmental Psychology,
1972, 6(2), 260-270.

Monkeys were reared during the first year of life (a) in a
feral environment and in the laboratory (b) with mothers
and peers, (c) with a surrogate mother or alone in a bare
wire cage; or in total social sensory isolation for (d) 6
months or (e) 9-12 months from birth. Following extensive
social and nonsocial experiences from years 1-4, the
monkeys were studied for motor activity and exploratory
behavior in a novel environment, and exploration of
visual stimuli varying in complexity. The experiments
were conducted in two independent replications, and

were run approximately 3 years apart. Both motor
activity and exploration were increasingly greater,
the greater the degree of stimulus variation present
in the rearing situation. Responsiveness to complex
visual stimuli was also higher for monkeys raised in
increasingly more complex environments. The effects
of deprivation rearing on exploratory behavior did,
however, depend critically on sex. Females were much
more active and exploratory than males after rearing
in partial or total isolation situations. These
results indicate that rearing experiences in monkeys
produce effects on exploration and visual complexity
preference that persist into adulthood, but some of
these effects occur primarily in males.

531. Sackett, G.P. Innate mechanisms in primate social behavior.
In C.R. Carpenter (Ed.), Behavioral regulators of behavior in
primates. Lewisburg, Pa.: Bucknell University Press, 1973, 56-67.

This paper describes the visual and auditory elicited fixed-
action patterns of infant rhesus monkeys (Macaca mulatta)
under total isolation conditions. The data presented in
this article indicates that visual stimulation by socially
meaningful events and subsequent unlearned responses are not
sufficient conditions for adequate socialization. These
events must be coupled with relevant informational feedback
if they are to have an impact on the social development of
the organism.

532. Sackett, G.P. Sex differences in rhesus monkeys following
varied rearing experiences. In R.C. Friedman, R.M. Richart, &
R.L. Vande Wiele (Eds.), Sex differences in behavior. New York:
Wiley, 1974, 99-122.

The studies reviewed in this paper deal with six basic
rearing conditions utilized in experiments with infant
rhesus monkeys (Macaca mulatta). The following conditions
are discussed: (1) total social isolation; (2) partial
social isolation; (3) surrogate mothering; (4) peer contact
but no mothering; (5) mothering and peer contact; and
(6) rearing in a natural environment (feral). The results
of these investigations are analyzed with regard to social
development in primates.

533. Sackett, G.P. Syndromes resulting from object deprivation--
2: Effects on development of rhesus monkeys. In J.H. Cullen (Ed.),
Experimental behavior: A basis for study of mental disturbance.
Dublin: Irish Univer-ity Press, 1974, 43-69.

> This paper describes some of the specific effects of
> rearing in total deprivation from social experience
> with varied conditions of deprivation from nonsocial
> object experiences. The two general types of deprivation
> rearing experiences which are discussed include the total
> and partial isolation.

534. Sackett, G.P. & Ruppenthal, G.C. Development of monkeys
after varied experiences during infancy. In S.A. Barnett (Ed.),
Ethology and development. London: Heinemann Medical Books, Ltd.,
1973, 52-87.

> This paper reviews studies of socially deprived rhesus
> monkeys (Macaca mulatta) and concomitant alterations in
> social, sexual, maternal, exploratory, and other behavior
> patterns. Innate behavior and fixed action patterns are
> also discussed with regard to primate ontogeny under
> feral and experimental conditions.

535. Sackett, G.P., & Ruppenthal, G.C. Some factors influencing
the attraction of adult female macaque monkeys to neonates. In
M. Lewis & L.A. Rosenblum (Eds.), The origins of behavior--The
effect of the infant on its caregiver. New York: John Wiley and
Sons, 1974.

> This article reviews a series of experiments concerned
> with identifying sources of maternal motivation. The
> subjects were monkey species that originated in
> either India (Macaca mulatta, the rhesus monkey) or
> the Malaysia-Indonesia Archipelago (Macaca nemestrina,
> the pigtail monkey). These studies measured preferences
> for neonates relative to other types of social stimuli,
> as indexed by approach and proximity of adult female
> subjects to live monkey stimuli. The factors studied
> for their potential influence in determining neonate
> attractiveness included (1) characteristics of infants
> as stimuli, (2) age and parity of adult female subjects,
> (3) mother-infant separation, (4) conditions under which
> the mother had reared her infant, and (5) conditions
> under which the mother herself had been raised.

536. Sackett, G.P., Holm, R., & Landesman-Dwyer, S. Vulnerability for abnormal development: Pregnancy outcomes and sex differences in macaque monkeys. In N.R. Ellis (Ed.), Aberrant development in infancy. Hillsdale, N.J.: Lawrence Erlbaum Associates, 1975, 59-76.

> This paper considers the possibility of a nonhuman
> primate model for the experimental study of the causes,
> consequences, and prevention of abnormal pregnancies,
> the primary goal of early postnatal factors related to
> abnormal or retarded behavioral development in infant
> primates. Considerations within this manuscript
> include: (1) human data on the causes and effects of
> poor pregnancy outcomes and sex related differences
> in susceptibility; (2) the results of an epidemiological
> study of pregnancies in pigtail monkeys raised in a
> colony situation; and (3) the differential effects of
> privation rearing conditions on male and female rhesus
> macaques.

537. Sackett, G.P., Holm, R.A., & Ruppenthal, G.C. Social isolation rearing: Species differences in behavior of macaque monkeys. Developmental Psychology, 1976, 12(4), 283-288.

> Social and nonsocial behaviors of infant rhesus
> (Macaca mulatta) and pigtail (M. nemestrina) monkeys
> reared in total social isolation were compared with
> those of socialized controls. As a species, pigtail
> monkeys were more social, more passive, and less
> exploratory. Rhesus isolates exhibited a "typical"
> syndrome of abnormal behavior to a much greater extent
> than pigtail isolates. These results question the
> generality of rhesus total isolate behavior as a
> model for some human problems and illustrate the need
> to study many species, even among closely related
> nonhuman primates.

538. Sackett, G.P., Porter, M., & Holmes, H. Choice behavior in rhesus monkeys: Effects of stimulation during the first month of life. Science, 1965, 147, 304-306.

> Monkeys reared from birth away from other monkeys and
> handled by humans during the first month of life
> preferred humans to monkeys when tested at the age
> of 2 to 3 years. Animals having both early human

handling and physical contact with other monkeys, or
physical contact with other monkeys and no human
handling, preferred monkeys. Subjects reared in
complete isolation from humans and monkeys spent
less time with either choice stimulus, but also
preferred monkeys to humans.

539. Sackett, G.P., Holm, R.A., Davis, A.E. & Fahrenbruch, C.E.
Prematurity and low birth weight in pigtail macques: Incidence,
prediction, and effects on infant development. Symposium, 5th
Congress of the International Primatological Society, 1974, 189-205.

The data reported in this article support the practicality
of a nonhuman primate model for the study of parental,
prenatal, and perinatal factors influencing postnatal
behavioral development. This research has demonstrated
that captive pigtail monkeys (Macaca nemestrina) produce
premature and low birth weight offspring at high rates.
A number of parental characteristics seem to be related
to high risk pregnancy outcome in specific breeding pairs.
Many of these parental factors (e.g., maternal and paternal
age, parity, and social stress) are also related to high
risk incidence in human pregnancies. The preliminary
weight, intake, and behavioral data also suggest that pigtail high risk newborns differ markedly from low risk
monkeys in early development.

540. Sackett, G.P., Holm, R.A., Ruppenthal, G.C., & Farhrenbruch,
C.E. The effects of total social isolation rearing on behavior
of rhesus and pigtail macaques. In R.N. Walsh and W.T. Greenough
(Eds.), Advances in behavioral biology, Vol. 17, Environments
as therapy for brain dysfunction. New York: Plenum Press, 1976.

The purpose of this article is to report comparisons
between rhesus and pigtail monkeys reared under identical
conditions of total social isolation. The results indicate
that rhesus monkeys, as a species, appear to be more
susceptible to isolation rearing effects than pigtails.
Although the direction of sex differences found between
isolates was generally the same, the magnitude of
differences was much greater for rhesus monkey isolates.

541.   Sackett, G.P., Bowman, R.E., Meyer, J.S., Tripp, R.L., &
Grady, S.A. Adrenocortical and behavioral reactions by differentially raised rhesus monkeys. Physiological Psychology, 1973, 1
(3), 209-212.

    Monkeys reared in social isolation or with an agemate
    were tested at 19 months for cortisol responses
    following (1) ACTH injection and (2) initial
    experience in a novel playroom. Isolates had higher
    basal cortisol, but did not differ from peer-raised
    monkeys in absolute cortisol rise induced by ACTH
    or by the playroom experience. However, isolates
    did show more fear-disturbance-emotional behavior
    than did peers. The data suggest that basal
    cortisol level may be a meaningful correlate of
    behavioral differences produced by differential rearing
    experiences in monkeys, but cortisol rises induced
    by novel and complex stimulation are not correlated
    with these behavioral effects.

542.   Sackett, G.P., Griffin, G.A., Pratt, C., Joslyn, W.D. &
Ruppenthal, G. Mother-infant and adult female choice behavior
in rhesus monkeys after various rearing experiences. Journal of
Comparative and Physiological Psychology, 1967, 63, 376-381.

    Adult female and infant monkeys reared under varying
    conditions chose between their own vs. other mothers
    or infants, unfamiliar adults vs. infants, and
    females varying in age. (a) Normally reared mother-
    infant pairs chose own mother or infant; multiple
    mothering produced no preference for own mother or
    infant; motherless mothers (MMs) did not prefer their
    own infants, although their infants preferred own
    mothers. (b) Normally reared infants preferred
    infants over adults; multiple-mothered infants
    preferred adults; infants of MMs strongly preferred
    an adult; all but normally reared mothers chose an
    unfamiliar adult over an infant. (c) Laboratory-born
    females preferred adults to younger monkeys regardless
    of previous births; feral-born females exhibited increased
    neonate preference with increased births.

543. Savage, E.S., Temmerlin, J., & Lemmon, W.B. The appearance
of mothering behavior toward a kitten by a human-reared chimpanzee.
Contemporary Primatology, Proceedings of the International Congress
of Primatology, Nagoya, 1974, 287-291. (Karger/Basel, 1975)

    A human-reared chimpanzee who had no previous experience
    with cats responded aggressively when first exposed to
    a domestic cat. Aggressive behavior was artificially
    inhibited and shortly thereafter become superseded by
    quasi-maternal behavior. Affiliative patterns first
    replaced aggressive patterns when the kitten began
    to spontaneously follow the chimpanzee. Even though
    the chimpanzee has had no contact with other chimpanzees,
    she displayed maternal behavior which was, insofar as
    the nature of the kitten would allow, consistent with
    that shown by other chimpanzee females toward their
    infants. This affiliative quasi-maternal behavior
    toward the kitten did not generalize to other cats.

544. Saxon, S.V. Differences in reactivity between asphyxial and
normal rhesus monkeys. The Journal of Genetic Psychology, 1961,
99, 283-287.

    Observations of young monkeys subjected to asphyxia
    neonatorum and normal control monkeys under four
    different stimulus situations showed consistent and
    significantly greater emotionality in the normal
    group as compared to that of the asphyxial group.
    No significant differences existed between groups
    in frequency of locomotion or in stimulus contact
    scores. Normal animals, however, generally appeared
    to be more responsive and more reactive to immediate
    environment than were asphyxial animals.

545. Saxon, S.V. Effects of asphyxia neonatorum on behavior in
the rhesus monkey. Journal of Genetic Psychology, 1961, 99, 277-
282.

    Systematic observations of preadolescent rhesus monkeys
    (Macaca mulatta) subjected to asphyxia neonatorum and
    normal control monkeys showed differences between groups
    in general reactivity to environmental surroundings.
    Asphyxia subjects locomoted significantly more, tended
    to contact objects more, and were significantly less
    emotional than normal subjects.

546. Saxon, S.V., & Ponce, C.G.   Behavioral defects in monkeys asphyxiated during birth. Experimental Neurology, 1961, 4, 460-469.

The behavior of five young rhesus monkeys subjected to asphyxia neonatorum and five normal control monkeys was studied. No significant differences between groups were found on object discrimination, delayed response, and perseveration tasks. Statistically significant differences were apparent in tests of sensori-motor capabilities (manual dexterity and patterned strings). In addition, the animals that had been asphyxiated were found to be consistently less emotional and less reactive than were normal control animals. A correlation with a previously described pattern of structural brain damage was drawn.

547. Schaeffer, H.H.   Self-injurous: behavior; shaping head banging in monkeys. Journal of Applied Behavior Analysis, 1970, 3(2), 111-116.

Head-banging, a common phenomenon among the mentally retarded, was shaped, brought under stimulus control, extinguished, and re-established in two monkeys through reinforcement and discrimination procedures of operant conditioning. The behavior was stable and led to lacerations, a condition that qualifies head-banging as self-injurious. The principles of the analysis of behavior used here may well be of value in the etiology and treatment of some human head-banging.

548. Schiff, W., Caviness, J.A., & Gibson, J.J.   Persistent fear responses in rhesus monkeys to the optical stimulus of "looming". Science, 1962, 136, 982-983.

The approach of an object corresponds with a spatiotemporal optical stimulus consisting of a symmetrical expansion of a closed contour in the field of view. The visual equivalent of impending collision was isolated and compared with its sequential inversion. Infant and adult rhesus monkeys manifested persistent avoidance responses to "looming" but not to the inverse. This visual stimulus alone is a strong exciter of avoidance, and the response appears early in life.

549. Schiltz, K.A., Thompson, C.I., Harlow, H.F., Mohr, D.J., &
Blomquist, A.J. Learning in monkeys after combined lesions in
frontal and anterior temporal lobes. Journal of Comparative and
Physiological Psychology, 1973, 83(2), 271-277.

    Nineteen monkeys divided into two unoperated control
    groups--a group sustaining prefrontal lobectomy and
    a group with prefrontal lobectomy plus removal of
    the anterior temporal neocortex--were compared on
    delayed response ability, as well as on the ability to
    make object discriminations and to form learning
    sets. The combined frontal-temporal lesion did not
    increase the delayed response deficit produced by
    frontal damage alone, but it significantly increased
    difficulty in making two-object discriminations and
    apparently also depressed oddity discrimination
    performance. Ability to form learning sets was not
    impaired in either lesioned group.

550. Schlottman, R.S., & Seay, B. Mother-infant separation in the
Java monkey (Macaca irus). Journal of Comparative and Physiological
Psychology, 1972, 79(2), 334-340.

    Twelve 7-8-mo.-old Macaca irus monkeys were separated from
    mothers and reunited. Infant playing and related behaviors
    decreased during the 3-wk. separation period. Presence
    of a conspecific unfamiliar adult female mother-substitute
    did not alleviate debilitating effects of separation.
    Effects were not attributable to traumatic process of
    forceful removal of mothers since a control group
    separated and immediately returned was minimally affected.
    During reunion, infants responded differtially to their
    own mothers regardless of whether they were returned to
    original or spatially reoriented home cages. Most
    behaviors returned to preseparation levels, but the
    mother-infant relationship was intensified in 3-wk.-
    separated animals. Species differences in reaction to
    maternal separation are discussed.

551. Scollary, P. Mother-infant separations in rhesus monkeys
(Macaca mulatta). (Doctoral dissertation, U.C., Davis, 1970).
Dissertation Abstracts International, 1971, 32(2).

A three phase reaction to separation has been described
for children: protest, despair and withdrawal. The
final phase occurs after reunion and involves either
psychological or physical rejection of the mother. In
previous studies on non-human primates only the
first two phases have been observed. Results of this
study on rhesus monkeys (M. mulatta) indicate that some
infants, in this case 4 males, do show this third stage
of rejection after reunion. This detachment phase was
observed again in the same animals after a second
separation.

552. Scott, J.P. Critical periods in behavioral development.
Science, 1962, 138, 949-958.

The article presents a very general analysis of
critical periods and their relationship to the
development of various organisms both psychologically
and physiologically. Primary processes of socialization,
affectional systems, learning, and innate mechanisms
are discussed in light of critical developmental
periods and the need for homeostatic stimulation during
this time.

553. Seay, B.M. Maternal behavior in primiparous and multiparous
rhesus monkeys (Doctoral dissertation, University of Wisconsin,
1964). Dissertation Abstracts International, 1965, 64-10, 305,
2629.

Two groups of four feral rhesus monkey mothers and
their infants were observed for six months. Primiparous
and multiparous mothers were strikingly similar on most
measures of positive maternal behavior as were their
infants on the basic infant-mother categories. Multiparous
mothers showed significantly more physical rejection than
primiparous mothers and there was nonsignificant differences
in the same direction for the categories of threatening and
punishment. The infants of the multiparous mothers
received significantly higher scores on the infant-mother
categories of approach and gross nonventral contact. The
data of the present experiment do not necessarily support
the conclusion that the maternal care given by primipara
is as fully adequate as that provided by multipara, but
there is no doubt that primiparous rhesus normally give
adequate care to their infants. The social behavior of

the infants of the primiparous and multiparous mothers
of the present experiment and of our surrogate raised
infants was considered with particular emphasis being
given to the development of sex differences and sexual
behavior. The obtained between-sex differences in
social behavior were not as clear as had been previously
reported. The discrepancy was attributable to individual
peculiarities of the male infants of the primiparous
mothered group. There as a suggestion of delay in the
appearance of complex sexual behaviors by surrogate reared
infants and some evidence of sexual specificity in
sexual behavior across all infant groups. The similarity
in the social behavior of surrogate and mother-reared
infants was attributed to multiple compensating bases
for the development of social behavior in the rhesus
monkey.

554. Seay, B.M. & Harlow, H.F. Maternal separation in the rhesus
monkey. Journal of Nervous and Mental Disease, 1965, 140(6), 434-
441.

The present study involved separation of eight monkey
mother-infant pairs for a period of two weeks and
measurement of the behavior of the infants before, during,
and after reunion with their mothers, and of the
mothers before and after reunion. All infants showed
emotional disturbance in response to separation and
drastic decreases in play and other complex social
behaviors while separated. It is clear that infant-
mother separation produces emotional disturbance in
both human and macaque infants and that the patterns
of responses following separation are similar in both
species. The results obtained in studies of monkey
infant-mother separation indicate that sheer physical
separation is the crucial aspect of maternal separation
for monkeys. Undoubtedly other factors associated with
separation from the mother are vitally important for
human children, and may account in part for the absence
or rarity of the detachment stage (as seen in human
response pattern) in separated monkey infants. The
overall results show considerable similarity in the
responses of human children and infant monkeys to
separation from the mother.

555. Seay, B., Alexander, B., & Harlow, H.F. Maternal behavior
of socially deprived rhesus monkeys. Journal of Abnormal and
Social Psychology, 1964, 69(4), 345-354.

The maternal behavior of 4 monkeys separated from their
mothers at birth and denied opportunity to interact
with other monkeys during the first 18 mo. of life
is described and compared with the maternal behavior of
4 normal feral monkeys. The 4 socially deprived
monkeys were grossly inadequate mothers, but the
social development of the infants of these inadequate
mothers was apparently normal. The maternal behavior
of 5 additional mother-deprived monkeys is also
described; 3 of these animals were inadequate and
2 adequate mothers. The 2 adequate mother-deprived
mothers had early social experience; the 7 inadequate
mothers did not. Three inadequate mothers have given birth
to a second infant and exhibited normal maternal care.

556. Seay, B., Hansen, E., & Harlow, H.F. Mother-infant separation
in monkeys. Journal of Child Psychology and Psychiatry, 1962,
3, 123-132.

The present study involved separation of four monkey
mother-infant pairs for a 3-week period and measurement
of the behaviour of the subjects before separation, during
separation, and after reunion. All mothers and all
infants showed emotional disturbance in response to
separation, but the infants' disturbances were more
intense and more enduring than those of the mothers.
The findings were interpreted as being generally in
accord with Bowlby's theory of primary separation
anxiety as an explanatory principle for the basic
primate separation mechanisms.

557. Sidowski, J.B. Altruism, helplessness, and distress:
Effects of physical restraint on the social and play behavior of
infant monkeys. Proceedings of the 78th Convention of the
American Psychological Association, 1970, 5, 233-234.

This research investigation discusses the influence of
physical restraint and total isolation on the social
and play behaviors of infant rhesus monkeys (Macaca mulatta)
over a six month period. Various behavior patterns were
observed, specifically altruistic, helpful, or sadistic

responses.

558. Sidowski, J.B. Psychopathological consequences of induced
social helplessness during infancy. In H.D. Kimmel (Ed.),
Experimental psychopathology: Recent research and theory. New
York: Academic Press, 1971, 231-249.

   This paper describes psychopathological behaviors resulting
   from the physical restraint of infant rhesus monkeys
   (Macaca mulatta) in dyadic social stiuations. Twelve
   rhesus monkeys were removed from their mothers at birth,
   placed in nursery care for two days, and then moved to
   individual cages enclosed with reflectionless, black
   plexiglass for six months. Physical restraint of one
   of a pair of animals provided an opportunity for
   either altruistic or sadistic behaviors on the part of
   the unrestrained conspecific. The results of this
   investigation point to the lack of altruistic behavior
   demonstrated by animals maintained in isolation conditions.

559. Sidowski, J.B., Harlow, H.F., & Suomi, S.J. Enhancing social attachment through fear: A study of infant monkeys. Psychonomic Science, 1972, 29(5), 323.

   For several months after birth, pairs of infant
   monkeys or infant/surrogate combinations were
   subjected to periodic presentations of a
   "monster" fear stimulus. The resultant emotional
   and social attachment effects will be discussed.

560. Sieler, C., Cullen, J.S., Zimmerman, J., and Reite, M.
Cardiac arrhythmias in infant pigtail monkeys following maternal
separation. Psychophysiology, 1979, 16(2), 130-135.

   The occurrence of nocturnal cardiac arrhythmias was
   determined in 9 infant pigtail (M. nemestrina) monkeys
   during 3 baseline normal nights, 3 nights following
   maternal separation, and 3 nights following reunion
   with the mother. All infants lived in social groups
   where they had been raised by their natural mothers;
   heart rate data were collected by means of totally
   implantable biotelemetry systems. Marked individual
   differences were found in the mean frequency of cardiac

arrhythmias in the baseline condition, and infants with
lower heart rates had a greater mean number of
arrhythmias. Maternal separation was accompanied by
both increases in arrhythmias and decreases in heart
rate. The relationship (slope of the regression line)
between arrhythmias and heart rate changed in 8 of the 9
infants during separation, suggesting that the increase
in arrhythmias was greater than could be accounted for
by the decreases in heart rate alone. While both heart
rate and arrhythmia values tended to return to baseline
values following reunion with the mother, some infants
exhibited prolonged separation-induced alterations.

561. Simons, R.C., Bobbitt, R.A., & Jensen, G.P. An experimental
study of mother M. nemestrina's responses to infant vocalizations.
American Zoologist, 1967, 7. (abstract No. 112)

This study is based on a catalog of vocal repertoires of
non-human primates, and is an experimental test of
the response of mother monkeys to two infant calls.
Two different vocalizations from each of two infants
and a control vocalization from an adult female were
tape-recorded and played to mother monkeys separated
from their infants and isolated in a sound-proof room
to adult female non-mother monkeys and to adult male
monkeys. In this situation mother monkeys, and only
mother monkeys, responded regularly and consistently
to the stimuli with increased pacing and calling.

562. Simons, R.C., Bobbit, R.A., & Jensen, G.D. Mother monkeys'
(Macaca nemestrina) responses to infant volcalizations. Perceptual
and Motor Skills, 1968, 27, 3-10.

Two mother monkeys (Macaca nemestrina) separated from
their infants, 2 adult females without young and 2 adult
males were studied to determine activity (pacing) and
vocalization in response to taped monkey calls. Stimulus
tapes were prepared from two different calls of each of the
two infant monkeys and a call of an adult female monkey.
Mother monkeys separated from their infants were always
more active and vocal than either non-mother female
monkeys or males, and their activity and response
vocalizations increased during presentations of infant
calls. There was no evidence that mothers responded
differentially to the calls of their own infants.

563. Singh, M. Mother-infant separation in rhesus monkeys living
in natural environment. Primates, 1975, 16(4), 471-476.

      Four mothers were separated from their infants in two
      free-ranging groups of rhesus monkeys. Infants were
      observed for three stages i.e. pre-separation,
      separation, and post-separation. Separation caused a
      marked decrease in play in the infant. Crying and
      restlessness increased. During post-separation, a
      significant increase in mother's approach behavior
      towards the infant was observed. The results of
      the present field study almost resemble the
      results of laboratory studies done by Seay,
      Hansen, and Harlow (1962).

564. Singh, S.D. Effects of infant-infant separation of young
monkeys in a free-ranging natural environment. Primates, 1977,
18(1), 205-214.

      Five infant rhesus monkeys reared together in a free-
      ranging natural environment for about eight months
      were subjected to peer separation. Separation of an
      infant was inflicted by way of removing all the other
      infants from their home range. The infants so separated
      showed exceedingly high levels of locomotion and
      vocalizations in the beginning of the separation phase, but
      their such excited behaviour did not last for more than
      a day, and during the remaining separation period they
      appeared to be quite depressed as mainly indicated
      by their reduced range of locomotion, reduced motivation
      for food and water, and reduced level of vocalization
      and environmental exploration. Thus, the animals
      showed a biphasic response to separation, which was
      characterized by an initial phase of 'protest' followed
      by a 'despair' stage, basically similar to what has been
      reported in infant rhesus monkeys subjected to peer or
      mother separation under laboratory conditions.

565. Spencer-Booth, Y. The behaviour of group companions towards
rhesus monkey infants. Animal Behaviour, 1968, 16, 541-557.

      The maternal behaviour of male and female group companions
      towards rhesus monkey (Macaca mulatta) infants was
      analyzed according to age of group companions, and in
      the case of females whether or not they have borne live

young and whether or not these were present. The most
conspicuous differences between categories of females
were that those which had live young showed less
behaviour than those which had not, and a lower
proportion of them did so. The proportions of the
different types of behaviour also differed most between
those groups of categories. Females about 2 years old
were most likely to show behaviour. Two-year-old males
were more likely to show behaviour than those younger or
older. Females showed more behaviour, and a higher
proportion were involved, than males of an equivalent
age class. Siblings showed more behaviour towards an
infant than an animal of the same category also
present in the group. When the mother was removed from
the group, the categories tended to stay in the same
relation to one another as when she was present, and the
amount of cuddles, hits, and grooming received by the
infant increased.

566. Spencer-Booth, Y., & Hinde, R.A. The effects of separating
rhesus monkey infants from their mothers for six days. Journal
of Child Psychology and Psychiatry, 1967, 7, 179-197.

This study is concerned with the effects of a 6-day
period of maternal deprivation on the behaviour of
four 30-32-week-old rhesus monkeys (Macaca mulatta).
The animals lived in a complex group situation, and
various aspects of mother-infant interaction and
infant behavior before, during and after the separation
period were observed and recorded. Data suggest that
the effects of separation shown by these infants are
similar to those seen in human children who have
had the same sort of experience.

567. Spencer-Booth, Y. & Hinde, R.A. Effects of brief separations
from mothers during infancy on behaviour of rhesus monkeys 6-24
months later. Journal of Child Psychology and Psychiatry, 1971,
12, 157-172.

Data on mother-infant relations, activity, and behaviour
in test situations were obtained from rhesus monkey
(Macaca mulatta) infants, age 12 and 30 months, which
had one or two 6-day separation experiences when 20-32
weeks old, and from non-separated controls. At 12 months
the previously separated infants showed less locomotor

activity than controls, and some difference from them
in mother-infant relations. There were no significant
differences in their readiness to approach strange
objects in the home pen, but marked differences when
they were tested in a strange cage. At 24 months
some differences were still present.

568. Spencer-Booth, Y., & Hinde, R.A. Effects of 6 days separation from mother on 18- to 32- week old rhesus monkeys. Animal
Behavior, 1971, 19, 174-191.

Nineteen group-living infants were subjected to separation
when 18, 21, 25 to 26 or 30 to 32 weeks old. The
21 and 25 to 26 week groups were separated again at 30 to
31 weeks. Effects on various activity measures and
the mother-infant relationship were assessed
quantitatively. During separation activity and play
decreased. When the mother returned the infants were more
clinging than before separation. Some of the effects
of separation were still apparent 4 weeks later. The
sex of the infants, contact with group companions
during separation and age at separation did not account
for variability in the response to separation. Infants
separated for the second time responded similarly to
infants separated for the first time at the same age.

569. Spencer-Booth, Y., & Hinde, R.A. The effects of 13 days
maternal separation on infant rhesus monkeys compared with those
of shorter and repeated separations. Animal Behavior, 1971, 19,
595-605.

The effects of removing the mothers of rhesus monkey
infants for 13 days were qualitatively similar to those
found previously, when the mothers were removed for
6 days. The infants' behaviour changed markedly in the
first few days of separation; thereafter there was no
clear evidence, from the measures used, of any further
changes. After the mothers' return these infants gave
more distress calls, and showed a greater depression of
locomotor and play behaviour, than did the infants whose
mothers had been removed for only 6 days.

570. Stevens, C.W., & Mitchell, G. Birth order effects, sex differences, and sex preferences in the peer-directed behavior of rhesus infants. International Journal of Psychobiology, 1972, 2 (2), 117-128.

    Eight firstborn rhesus monkey infants (Macaca mulatta) were compared to eight later-born infants with regard to their social-emotional adjustment. All subjects were paired at 6 to 11 months of age with six sex-balanced stimulus strangers (rhesus infants of similar age). Firstborn infants vocalized more frequently, were more active, and displayed more stereotyped movements than did later-born infants. The firstborns were also more assertive and sociable and were in the proximity of the stimulus infants more frequently than were the later-born infants. Thus the development of social-emotional behavior in the firstborn monkeys does not appear to be seriously debilitated during infancy by anxious, inexperienced mothering and/or by longer, more difficult births. Sex differences in infant rhesus monkey behavior were found which resembled those reported elsewhere, but sex preference data reported elsewhere were not corroborated.

571. Strobel, D.A. Stimulus change and attentional variables as factors in the behavioral deficiency of malnourished developing monkeys (Macaca mulatta) (Doctoral dissertation, University of Montana, 1972). Dissertation Abstracts International, 1973, 33(11).

    Five experiments tested, over a two year period, four groups of monkeys subjected to postweaning diets that were either radically deficient or enriched in protein content. The studies, in general, were designed to evaluate the supposition that protein diets affected perceptual and motivational factors governing stimulus control of specific behaviors.

572. Strobel, D.A. Protein malnutrition in monkeys. In K.F. Riegel and J.A. Meacham (Eds.), The developing individual in a changing world. Vol. 2. Chicago: Aldine, 1976, 465-473.

    The present experimental program with developing monkeys (M. mulatta) was designed to evaluate changes in chronically malnourished animals during the interval between weaning and sexual maturity. These behaviors were then re-examined after rehabilitation to determine

if adult or maturing behavior was altered. The
value of a non-human primate model for human malnutrition is discussed.

573. Strobel, D.A., & Zimmermann, R.R. Manipulatory responsiveness in protein-malnourished monkeys. Psychonomic Science, 1971,
24(1), 19-20.

Two groups of rhesus monkeys were placed on diets
containing either 3% or 25% protein at 210 days
of age and tested consecutively on two puzzle
manipulation tasks varying in difficulty. Malnourished
Ss were able to learn to solve the puzzles in the
absence of food rewards but showed significantly
lower final performance levels on the easier puzzles
than did controls. The more complex puzzles were
learned with equal difficulty by the high- and low-
protein groups. Manipulatory responsiveness in this
study was interpreted to be partially influenced by
secondary diseases associated with protein-calorie
malnutrition.

574. Strobel, D.A., & Zimmernann, R.R. Responsiveness of protein
deficient monkeys to manipulative stimuli. Developmental Psychobiology, 1972, 5, 291-296.

Objects never experienced before by 4 groups of
monkeys were introduced into a free operant chain
manipulation situation. The presence of the objects
disrupted the manipulation response rates of 2 groups
of animals differing in age and onset of a diet
restricted in protein. Age controls receiving higher
concentrations of protein in diets isocaloric with
respect to the protein restricted diet, showed
increased performance to these stimuli. The reinstate-
ment of the chains alone condition resulted in a partial,
but temporary recovery, of the manipulation response by
low protein animals. This disruption in chain pulling
behavior by malnourished monkeys may be similar to
neophobic avoidance reactions previously observed in
protein deprived monkeys.

575.   Strobel, D.A., Geist, C.R., Zimmermann, R.R. & Lindvig, E.K.
Cue-locus--a factor in the behavioral deficiency of the developing
protein-malnourished monkey (<u>Macaca mulatta</u>). <u>Behavioral Biology</u>,
1974, <u>10</u>, 473-484.

   Four groups of infant rhesus monkeys were separated
   from their mothers at 90 days of age, housed individually,
   and placed on purified diets which contained either 25
   or 3.5% protein at 120 or 210 days of age. At approximately
   1,375 days of age, brightness discrimination problems
   were presented in which the locus of the cues was either
   central or peripheral and the area of the different
   cues was varied. No significant differences were found
   for diet, cue area, or diet X area interaction for
   peripherally located cues on original learning, shift 1,
   or reversal conditions. A significant decline in
   performance levels was detected for all groups as the
   area of the cues decreased in the central locus condition.
   Subjects maintained on diets deficient in protein, which
   were reversed to central cues, exhibited a marked
   decrement in performance when compared to the high
   protein controls. Further, malnourished subjects were
   considerably inferior to high protein animals on the
   smallest cue areas in shift 1 and shift 2 leading to a
   significant diet X cue area interaction. Under reversal
   conditions for centrally located cues, low protein subjects
   were found to perform at significantly lower levels than
   the high protein controls. These results are consistent
   with both a visual sampling model of attention and a
   number of two-stage models of attention.

576.   Stoffer, G.R. & Stoffer, J.E.   Stress and aversive behavior
in non-human primates: A retrospective bibliography (1914-1974)
indexed by type of primate, aversive event, and topical area.
<u>Primates</u>, 1976, <u>17</u>(4), 547-578.

   A retrospective bibliography (1914-1974) of stress
   and aversive behavior in non-human primates is
   presented. Each of 582 citations is indexed for
   type of primate(s) studied, kind of aversive event(s),
   and topical area(s). The aversive events include
   air (pressurized and wind), alien species (human,
   human staring, and snake), aversive drugs, crowding,
   darkness, electrical stimulation of the brain,
   gravitational forces, hitting, isolation, looming,
   noise, nonreward (frustration), pinching, pricking,
   radiation, restraint, sandpaper, sensory deprivation,

shock (electric), slapping, social defeat (induced
experimentally), strobe light, tastes (foul), temperature
extremes, threat of social separation, time-out from
positive reinforcement, unavoidable aversive events,
unpredictable stimulus changes, vertical chamber confinement,
visual cliff, and water (rain). Index terms include
traditional and specific topics such as fear, punishment,
and complex schedules of negative reinforcement, topics
sub-divided within the areas of social behavior, ontogeny,
perceptual processes, and psychopathology, and topics
of special interest including space research, field
research, research subjects, research instrumentation,
shaping and training techniques, selected drugs (e.g.,
morphine), and cardiovascular or gastointestinal correlates
of stress. References whose primary interest were physiology or the study of drug effects were not included. In
no case was a citation included if there was no recording
of behavior.

577. Stuble, R.G. & Riesen, A.H. Changes in cortical dendrititc
branching subsequent to partial social isolation in stumptailed
monkeys. Developmental Psychobiology, 1978, 11(5), 479-486.

Stumptailed monkeys were reared from 1 week after
birth to 6 months of age in either a colony
condition with the mother or in partial social
isolation that allowed visual contact with the
colony animals, but not physical contact. At
6 months of age the animals were killed and
selected areas of the neocortex stained by the
Golgi-Cox mehtod. Relatively nonspiney cells
of Layer IV were drawn and analyzed for complexity
of dendritic brancing. Isolation-reared animals
had significantly decreased branching complexity
in Motor I cortex when compared to the control
animals. A transformation of the data that related
the number of branches to the number of previous
branches showed a slight rearing effect in
Somatosensory I cortex with the deprived animals
having a lower rate of branching than the controls.
We conclude that social isolation also includes a
motoric deprivation that could account for these
data.

578. Suomi, S.J. Factors affecting responses to social separation in rhesus monkeys. In G. Serban and A. Kling (Eds.), Animal models in human psychobiology. New York: Plenum Press, 1976, 9-26.

This article discusses the role of nonhuman primate research in the investigation of human psychopathology. Several investigations are discussed with a variety of macaque species and the variables affecting responses to social separation.

579. Suomi, S.J. Social development of rhesus monkeys reared in an enriched laboratory environment. The 29th International Congress of Psychology. Tokyo: Sasaki Co., 1972, 238.

Social development of rhesus monkeys raised in a nuclear family situation containing their parents, siblings, and other adult pairs and their offspring was studied at the University of Wisconsin Primate Laboratory. It was found that maternal initiations and offspring initiations and rejections but not maternal rejections differed according to infant birth order. Adult male-infant behavior varied as a function of infant sex. Infants preferred mothers, fathers, and siblings to other adult females, males, and peers, respectively. Infants exhibited more rapid and sophisticated social development than monkeys reared in less complex laboratory environments. The use of the nuclear family situations as a controlled rearing environment is discussed.

580. Suomi, S.J. Repetitive peer separation of young monkeys: Effects of vertical chamber confinement during separations. Journal of Abnormal Psychology, 1973, 81(1), 1-10.

Four rhesus monkeys were reared together, without mothers, for the first 3 months of life, then subjected to 20 peer separations and reunions during the next 6 months, being confined to vertical chambers during the periods of separation. Their behaviors were compared with those of four monkeys who were reared together without separations and four monkeys from a previous study who had been subjected to similar separations but not confined to vertical chambers. It was found that both groups of separated monkeys exhibited protest-despair reactions to the separations and in comparison with the non-

separated controls showed maturational arrest. However,
the chamber-separated subjects exhibited higher
levels of self-directed activity and lower levels of
socially directed activity than their cage-separated
counterparts. The significance of these findings with
respect to monkey models of human anaclitic depression
is discussed.

581. Suomi, S.J. Surrogate rehabilitation of monkeys reared in
total social isolation. Journal of Child Psychology and Psychiatry,
1973, 14, 71-77.

Four rhesus monkeys were reared in total social isolation
for the first 6 months of life, then exposed to surrogates
and subsequently housed in pairs. The isolates
exhibited significant decreases in abnormal behavior
patterns and developed crude patterns of social interaction
throughout this period, but failed to develop sophisticated social behaviors such as play. The rehabilitative
capabilities and limitations of surrogates for isolate-
reared subjects are delineated and discussed.

582. Suomi, S.J. Social interactions of monkeys reared in a
nuclear family environment vs. monkeys reared with mothers and
peers. Primates, 1974, 15(4), 311-320.

Four-year-old laboratory-born rhesus monkeys that had
been reared in a nuclear family social environment
consisting of mothers, fathers, siblings, peers, and
other adults of both sexes were permitted to interact
in various combinations with equal-aged monkeys that
had been reared in an environment consisting of only
mothers and peers. It was found that in most interaction
sessions nuclear family subjects exhibited significantly
higher levels of dominance and activity behaviors and
significantly lower levels of submissive and passive
behaviors than the mother-peer-reared subjects. These
differences were not evident when subjects were tested
within their own rearing groups. The significance of
the results with respect to previous and future studies
of social development in differential social environments
is discussed.

583. Suomi, S.J. Factors affecting responses to social separation
in rhesus monkeys. In G. Serban and A. Kling (Eds.), <u>Animal models</u>
<u>in human psychobiology</u> (Kittay Scientific Foundation International
Symposium, 1974). New York: Plenum Press, 1976, 9-26.

> This chapter presents data relevant to two forms of
> psychopathology in monkeys: that which results from
> social isolation from birth, and that which results from
> the breaking of attachment bonds under certain conditions.
> These forms differ significantly from each other in origin,
> in distinctive behavioral concomitants, and in
> procedures which successfully reverse the psychopathologies.
> The author believes that the first probably does not have
> a common human analog, while the second most likely
> does, particularly since it was initially based almost
> exclusively on human data. The results of a series of
> studies with monkeys are discussed with respect to
> their relevance to an animal model of human depression.

584. Suomi, S.J. Mechanisms underlying social development: A
reexamination of mother-infant interactions in monkeys. In A.D.
Pick (Ed.), <u>Minnesota Symposia on Child Psychology</u>, Vol. 10,
Minneapolis: University of Minnesota Press, 1976, 201-228.

> This chapter reviews the monkey data in the area
> of the development of mother-infant relationships, and
> presents some preliminary findings from a recently
> completed study. In addition, some alternative
> procedures are proposed that may produce monkey data
> more useful to human investigators than earlier
> methods have been.

585. Suomi, S.J. Development of attachment and other social
behaviors in rhesus monkeys. In T. Alloway, P. Pliner, and L.
Krames (Eds.), <u>Advances in the study of communication and affect</u>:
<u>Attachment behavior, Vol. 3,</u> New York: Plenum Press, 1977,
197-224.

> This chapter describes much of the recent research in
> social development of rhesus monkeys (<u>M. mulatta</u>) at the
> Wisconsin Primate Laboratory. This work has been oriented
> about three issues: what is the nature of a monkey's
> social capabilities at various points throughout the
> course of maturation, in what ways and to what degree can
> development be influenced by the nature of the social

environment, and, finally, what are the specific
mechanisms underlying the interaction between
response systems and social environment?

586.   Suomi, S.J.   Neglect and abuse of infants by rhesus monkey
mothers.   Voices, 1977, 12(4), 5-8.

   This article summarizes the research undertaken at
   the University of Wisconsin over the past ten years
   on "motherless mother" monkeys and their offspring.
   One of the most clear-cut findings to emerge from a
   recently concluded survey of 50 maternally deprived
   female monkeys and their infants, is that maternal
   competence was closely related to early socialization.
   A second consistent finding was that male offspring
   were far less likely to receive adequate maternal
   care and far more likely to be abused than were
   female offspring. A third finding was that as the
   incidence of adequate maternal behavior rose, the
   incidence of indifferent maternal behavior fell,
   and the incidence of abusive behavior remained
   relatively constant when motherless mothers had
   their second, third, fourth, and in a few cases,
   fifth offspring. Although it is not difficult
   to find suggestive parallels to these results in
   the human child abuse literature, the author
   cautions against using these findings, by themselves,
   as a basis for a valid and/or useful monkey model
   of human child abuse.

587.   Suomi, S.J., & Harlow, H.F.   Apparatus conceptualization for
psychopathological research in monkeys.   Behavior Research Methods
and Instrumentation, 1969, 1(7), 247-250.

   Investigation of psychopathology in monkeys requires
   analysis of multiple variables mediating specific
   behaviors. Assessment of both the relative contribution
   of each variable and the interactions among them is
   facilitated by the use of apparata whose designs
   permit separate or simultaneous manipulation of several
   variables. Three such devices, the pit, the tunnel
   of terror, and the standard living-experimental cage,
   each specifically constructed for the production of
   depression or despair in monkeys are described.

588. Suomi, S.J., & Harlow, H.F.   Abnormal social behavior in
young monkeys.   In J. Hellmuth (Ed.), The exceptional infant, Vol.
II. New York: Brunner/Mazel, 1971.

This chapter presents data generated from laboratory
studies of rhesus monkeys.   These studies have dealt
with normative development of rhesus monkey social
behavior, procedures by which aberrations in such
normative development can be reliably induced, and
procedures by which some of these abberations can be
reversed.

589. Suomi, S.J. & Harlow, H.F.   Monkeys at play.   Natural History,
1971, 80, 72-76.

This article presents a very general discussion of the
importance of play behavior in both human and nonhuman
primates.   Various studies are discussed with particular
emphasis on the development of these behaviors in infant
rhesus monkeys (Macaca mulatta) reared under an array of
social conditions.

590. Suomi, S.J., and Harlow, H.F.   Social rehabilitation of isolate-
reared monkeys.   Developmental Psychology, 1972, 6, 487-496.

Numerous researchers have indicated that 6 or more
months of total social isolation initiated at birth
produces profound and apparently permanent social deficits
in rhesus monkey subjects.   The present experiment successfully rehabilitated monkeys that spent the first 6 months
of life in total social isolation.   Following removal from
isolation subjects were permitted to interact with socially
normal monkeys 3 months chronologically younger than themselves.   Within a few weeks isolate disturbance behaviors
decreased substantially and were replaced by elementary
socially direct activity.   After 6 months of such exposure,
the isolate subjects were virtually indistinguishable from
their younger controls both in terms of absence of
disturbance behaviors and sophistication of social behaviors.

591. Suomi, S.J., & Harlow, H.F.   Depressive behavior in young
monkeys subjected to vertical chamber confinement.   Journal of Comparative and Physiological Psychology, 1972, 180(1), 11-18.

Young rhesus monkeys were confined in vertical
chambers for 6 wk., and their subsequent behavior
over a 9-mo. period in both a home-cage and playroom
situation was compared with that of like-aged
monkeys housed individually or in pairs. In comparison
to both control groups, chambered monkeys exhibited
excessive amounts of self-clasp and huddle, abnormally
low levels of locomotion and exploration, and an
absence of social interaction with other monkeys.

592. Suomi, S.J. & Harlow, H.F. Early experience and induced psychopathology in rhesus monkeys. Revista Latinoamericana de Psicologia, 1975, 7(2), 205-229.

A comprehensive review is presented of research at the
University of Wisconsin Primate Laboratory with emphasis
given to total social isolation. Unadjusted behaviors are
studied, strategies for correction are explored, and
explanations are given of the behaviors observed when
depriving the young monkey of social contact.

593. Suomi, S.J. & Harlow, H.F. Effects of differential removal
from group on social development of rhesus monkeys. Journal of
Child Psychology and Psychiatry, 1975, 16, 149-164.

Rhesus monkey infants were reared with surrogates in
groups of four for the first 4 months of life, then were
individually removed from their group for 28 days during the
succeeding 6-month period. Members of two groups were placed
in vertical chambers during their period of group separation,
while members of a third group were housed in single cages
during their time of separation. A control group of four
monkeys remained intact throughout the study. Results
indicated that following removal from group, subjects showed
lower levels of locomotion and play and higher levels
of clinging and self-clasping than nonseparated controls.
Differences from controls were exaggerated in the monkeys
confined in vertical chambers during separation. All
separation groups were socially inferior to the control
group by the end of the study.

595. Suomi, S.J. & Harlow, H.F.   The role and reason of peer relationships in rhesus monkeys. In M. Lewis & L.A. Rosenblum (Eds.), Friendship and peer relations. New York: John Wiley & Sons, 1975, 153-185.

> The development of peer interactions in rhesus monkeys (Macaca mulatta) are discussed with particular emphasis on the ontology of peer affection under feral and socially complex laboratory conditions.

596. Suomi, S.J., & Harlow, H.F.   The facts and functions of fear. In M. Zuckerman and C.D. Spielberger (Eds.), Emotions and anxiety: New concepts, methods, and applications. Hillsdale, N.J.: Lawrence Erlbaum Associates, 1976, 3-34.

> This chapter discusses the facts and functions of fear using data accumulated from years of laboratory study utilizing rhesus monkeys (M. mulatta) as subjects. It is the authors' contention that fear in monkeys, and most likely in all other primates, is an unlearned response system which soon after emergence into an animal's behavioral repertoire becomes subject to the contingencies of the environment. Further, it is their position that the fear response system has evolved over millions of years and represents a mechanism adaptive in at least two respects: it enhances the probability of survival of individual species members, and it promotes and maintains social structures within the species. A workable definition of fear as exhibited by rhesus monkeys is proposed, its chronological development traced, and experimental factors which affect its frequency and intensity of occurrence are described.

597. Suomi, S.J., & Harlow, H.F.   Monkeys without play. In J. Bruner, A. Jolly and K. Sylva (Eds.), Play--its role in development and evolution. New York: Basic Books, 1976, 490-495.

> This article discusses the role of play in the social development of infant monkeys. The authors believe that play among monkey infants serves two general, but important, functions. First, it provides a behavioral mechanism by which activities appropriate for adult social functioning can be initiated, integrated,

and perfected. Second, it mitigates aggression
when it emerges in the monkey's behavioral repertoire.

598. Suomi, S.J., & Harlow, H.F. Early separation and behavioral
maturation. In A. Oliverio (Ed.), Genetics, environment and
intelligence. Amsterdam: Elsevier/North-Holland Biomedical Press,
1977, 197-214.

This chapter presents an account of social maturation
in the rhesus monkey (Macaca mulatta). Normative
patterns of social development in the species are
described, and how such patterns can be influenced and
altered by various environmental contingencies are
discussed. The authors argue that separate theoretical
approaches are necessary to account for the specific
phenomena associated with two different classes of
heredity-environment interactions: those that are
associated with denying a monkey access to certain
classes of social stimuli and those that are
associated with the interuption of previously
established social relationships.

599. Suomi, S.J., & Harlow, H.F. Production and alleviation of
depressive behaviors in monkeys. In J. Maser and M.E.P. Seligman
(Eds.), Psychopathology: Experimental models. San Francisco:
W.H. Freeman, 1977, 131-173.

This chapter reviews research with monkeys which
pertains to the production and alleviation of
depressive behaviors. Effective criteria for
determining the validity and utility of animal
studies of human disorders are presented, and animal
data are presented within this context.

600. Suomi, S.J., & Harlow, H. F. Early experience and social
development in rhesus monkeys. In M.E. Lamb (Ed.), Social and
personality development. New York: Holt, Rinehart and Winston,
1978, 252-271.

This chapter discusses three major principles of
social development in monkeys. First, a number of
response systems emerge relatively independent of
an individual's rearing environment. Second, the
manner in which such response systems are integrated

into an individual's behavioral repertoire is highly
dependent upon the nature of the individual's past
and present social environment. Finally, although
the effects of various early social environments on
later behavior are clearly profound and persistent,
such effects are reversible. These points are explored
and supporting data presented.

601. Suomi, S.J., & White, L.E. Summary of "Early Social Behavior."
In D. Chivers and J. Herbert (Eds.), Recent advances in primatology, Vol. 1. London: Academic Press, 1978.

This summary emphasizes the role of nonhuman primate
mother-infant research on the recent advances in the
study of early social behavior in monkeys. Although
less than a decade ago "early social behavior" was
synonomous with "mother-infant interactions", investigators
have recently begun to seriously examine the developing
primate's social relations with conspecifics other than
its mother. The present volume reflects this new emphasis.

602. Suomi, S.J., Collins, M.L., & Harlow, H.F. Effects of
permanent separation from mother on infant monkey. Developmental
Psychology, 1973, 9(3), 376-384.

Rhesus monkey infants were reared from birth with their
mothers in a common pen, then separated permanently
from their mothers at 60, 90 or 120 days of age. Half
of the subjects in each age group were housed individually
following separation; the other half were housed in pairs.
Although the subjects did not differ appreciably in levels
of behaviors scored prior to separation, their behaviors
both in the week following separation and at six months
of age varied according to age of separation and
subsequent housing condition manipulations. All monkeys
showed agitation immediately following separation.
Monkeys separated at 90 days of age showed quantitatively
more severe immediate reaction to separation but by six
months of age did not differ appreciably from monkeys
separated at other ages. Monkeys housed singly following
separation showed significantly higher levels of
disturbance and self-clasping behavior and lower levels of
locomotion during the week following separation than
monkeys housed in pairs. These differences were exaggerated
at six months of age.

603.   Suomi, S.J., Collins, M.L., Harlow, H.F., & Ruppenthal, G.C.
Effects of maternal and peer separations on young monkeys.   Journal
of Child Psychology and Psychiatry, 1976, 17(2), 101-112.

> Rhesus monkeys reared with mothers and peers for the
> first 6 months of life were subjected to three
> consecutive 2-week periods of social separations,
> progressively increasing the degree of social isolation
> for the infants.   During each period of social separation--
> mesh permitting physical contact with mothers, physical
> but not visual separation from mothers but unlimited
> access to peers, and finally physical separation from
> both mothers and peers--clear patterns of protest and
> despair were recorded.   Reactions to mesh separations
> were mild relative to the latter two separations, both
> equally debilitating.   Following reunion, separated
> subjects returned to behavior patterns similar to
> those of nonseparated control monkeys.   Later in the
> study mothers were removed permanently from the home-
> cages of all subjects, and all subjects reacted with
> protest and cessation of play activity.   However, there
> was little evidence of despair and previously separated
> and control monkeys reacted in similar form despite their
> different early histories.

604.   Suomi, S.J., Delizio, R., & Harlow, H.F.   Social rehabilitation of separation-induced depressive disorders in monkeys.
American Journal of Psychiatry, 1976, 133(11), 1279-1285.

> The authors studied a group of four monkeys reared
> together, repeatedly separated from each other, and
> then exposed to another group of four monkeys reared
> in surrogate-peer groups who acted as therapists.   The
> study group was compared with the therapist monkeys,
> a group exposed to the same separations but not to
> the therapist monkeys, a control group that
> experienced no separations, and two additional
> groups of stimulus animals.   The authors' findings
> indicate that monkeys showing depressive behaviors
> after repeated separations can be returned to age-
> appropriate social performance through repeated
> exposure to socially active age-mates.

605. Suomi, S.J., Harlow, H.F., & Domek, C.J.   Effect of repetitive infant-infant separation of young monkeys. Journal of Abnormal
Psychology, 1970, 76(2), 161-172.

    Virtually all previous studies investigating the effects
    of separation from an attachment object have used a
    mother-infant model. In view of several methodological
    difficulties inherent in mother-infant separation, an
    alternative procedure, that of repetitive, short-term,
    infant-infant separations, is proposed. Results from
    three studies employing this procedure with young
    monkeys indicated that: (a) each short-term infant-infant
    separation produced behavior patterns similar to those
    exhibited by infants separated from their mothers; (b)
    Ss did not adapt to the separation procedure, but rather
    continued to show similar reactions to each successive
    separation; and (c) a cumulative effect of repetitive
    infant-infant separations was an arrest of maturation
    of social development in the monkey Ss. Implications
    of the findings are discussed.

606. Suomi, S.J., Harlow, H.F., & Kimball, S.D.   Behavioral effects
of prolonged partial social isolation in the rhesus monkey. Psychological Reports, 1971, 29, 1171-1177.

    Forty-eight monkeys, reared from birth in partial social
    isolation and ranging in age from 2 mo. to 13½ yr., and
    12 feral-born monkeys, comparable in age to the 12
    oldest partial isolates, were each scored for home-cage
    behavior. The survey showed that most partial isolates'
    behavioral levels declined with increasing age, males
    exhibiting more disturbance activity, more threats, and
    less grooming than females. The oldest partial isolates
    showed less locomotion and exploration but considerably
    more disturbance activity than their feral-born counterparts.
    The deficits produced by prolonged partial social isolation
    are discussed.

607. Suomi, S.J., Harlow, H.F., & Lewis, J.K.   Effects of bilateral
frontal lobectomy on social preferences of rhesus monkeys. Journal
of Comparative and Physiological Psychology, 1970, 70(3), 448-453.

Twelve rhesus monkeys (Macaca mulatta), 6 with bilateral
frontal lobectomies and 6 controls, were tested for
preference between a lobectomized and a nonlobectomized
stimulus animal of (a) the same sex as the subject, and
(b) the opposite sex from the subject. It was found that:
(a) when presented with stimulus animals of their own sex,
lobectomized subjects exhibited essentially no consistent
preference pattern, while control subjects preferred the
lobectomized stimulus animals, and (b) when presented
with stimulus animals of the opposite sex, lobectomized
subjects preferred the lobectomized stimulus animals,
while control subjects preferred the intact stimulus
animals. Implications of the findings are discussed.

608. Suomi, S.J., Harlow, H.F., & McKinney, W.T., Jr. Monkey
psychiatrists. American Journal of Psychiatry, 1972, 128(8),
927-932.

Most efforts that have been made to reverse the
effects of isolation on monkeys have been
unsuccessful. The authors report on successful
rehabilitation through the use of "therapist"
monkeys. The therapists, three months younger
than the isolate monkeys, initiated social
contact in a nonthreatening manner. Within
six months, the isolates' disturbance behaviors
had nearly disappeared and they displayed normal
age-appropriate social and play behaviors.

609. Suomi, S.J., Harlow, H.F., & Novak, M.A. Reversal of social
deficits produced by isolation rearing in monkeys. Journal of
Human Evolution, 1974, 3, 527-534.

A major body of research has clearly demonstrated
that rearing monkeys for at least the first 6 months
of life in total social isolation from conspecifics
produces profound psychopathology, including both
massive social deficits and idiosyncratic self-directed
behaviors rarely seen in normal monkeys. Previously,
isolation-induced psychopathology was thought to be
permanent, and various theoretical explanations, e.g.
critical periods and "emergence trauma" were posited
to account for such phenomena. Recent research at
Wisconsin has indicated that the isolation syndrome
in monkeys is, in fact, reversible, rendering the
above theoretical positions inadequate. In this

paper, the procedures for successful rehabilitation of
isolate-reared monkeys are described and an alternative
theoretical explanation presented. Implications of the
theory and data are discussed.

610. Suomi, S.J., Sackett, G.P., and Harlow, H.F. Development of
sex preference in rhesus monkeys. Developmental Psychology, 1970,
3, 326-336.

      Socially unsophisticated rhesus monkeys (Macaca mulatta)
      of various ages were tested for preference (a) between
      a male and a female stimulus animal the same age as the
      subjects, and (b) between a male and a female adult
      feral-born monkey. In sex preference for age-mates,
      (a) subjects of both sexes under 7 months of age failed
      to demonstrate consistent preferences for either sex of
      stimulus animal, (b) older subjects of both sexes
      made definite sex choices, exhibiting a chronological
      pattern of initial preference for own sex of stimulus
      animal followed by a shift of preference to opposite-
      sex stimulus animal, and (c) the preference shift
      occurred earlier for the female subjects than for
      males. The specific sex preferences exhibited by
      these unsophisticated subjects correspond closely to
      previously reported behavior patterns of socially
      sophisticated monkeys, and the chronological ages
      of preference "shifts" correspond to ages of physical
      maturation for rhesus monkeys. In sex preference for
      feral adults (a) all subjects under 38 months preferred
      the adult female to the adult male, suggesting the
      existence of an unlearned preference for adult
      females by infant monkeys; and (b) all males over
      40 months preferred the adult male, while all females
      over 40 months preferred the female.

611. Suomi, S.J., Eisele, C.D., Grady, S.A., & Harlow, H.F.
Depressive behavior in adult monkeys following separation from
family environment. Journal of Abnormal Psychology, 1975, 84(15),
576-578.

      Previous researchers have demonstrated that depressive
      behavior can be induced in infant monkeys via social
      separation procedures but that adult monkeys are
      relatively immune to similar manipulations. The
      present research studied the reactions of nuclear family-
      reared young adult rhesus monkeys to separation from

their families. Monkeys housed with friends during family
separation were relatively unaffected by the separation,
as were monkeys housed with both friends and strangers.
However, subjects individually housed following family
removal exhibited depressive-like behaviors previously
observed only in infant monkeys separated from mother
and/or peers. Implications of the findings are briefly
discussed.

612. Suomi, S.J., Eisele, C.D., Grady, S.A., & Tripp, R.L. Social
preferences of monkeys reared in an enriched laboratory environment.
Child Development, 1973, 44, 451-460.

Rhesus monkeys ranging in age from 1 month to 4 years
who had been reared in a nuclear family environment
with continual access to their mothers, fathers, and
siblings and with limited access to other parents and
their offspring were tested for preference (a) among
their mothers, familiar adult females, and unfamiliar
adult females; (b) among their fathers, familiar adult
males, and unfamiliar adult males; (c) among their
siblings, familiar peers, and unfamiliar peers; and
(d) between their mothers and fathers. It was found
that subjects of all ages preferred their mothers to
other adult females, preferred their fathers to other
adult males, preferred their mothers to their fathers,
but exhibited no preference among siblings and peers.
The implications of these findings are discussed.

613. Suomi, S.J., Seaman, S.F., Lewis, J.K., DeLizio, R.D., &
McKinney, W.T., Jr. Effects of imipramine treatment on separation-
induced social disorders in rhesus monkeys. Archives of General
Psychiatry, 1978, 35, 321-325.

Two groups of young rhesus monkeys were subjected to
repetitive peer separations, a procedure that has been
shown to produce depressive-like reactions in infant monkeys.
Midway through the procedure one group was treated with
the antidepressant imipramine hydrochloride, the other
with a saline placebo. In comparison with placebo
treatment, the imipramine treatment yielded significant
behavioral improvement in a form and with a time course
similar to that seen when the drug is given clinically
to human depressives. We discuss the implications of
the findings.

614. Symmes, D., Eisengart, M.A., & Healy, M.E.   Neurobehavioral studies of isolation-reared monkeys. Proceedings of the American Psychological Association, 1971, 6, 789.

   Data are presented on neurophysiological and biochemical
   correlates of abnormal behavior in isolation-reared rhesus
   monkeys. Two groups of four monkeys have been studied
   over a period of 6-24 mo. of age after differential
   rearing for the first 6 mo. of life. The experimental
   animals were separated from their mothers shortly after
   birth and reared with no physical contact with other
   monkeys and with minimum contact with humans. The
   control animals were raised in the laboratory in
   similar wire cages with their mothers. Behavioral
   differences between the groups were quantified, using
   videotape analysis under a variety of conditions
   including a 14-day social experience with the other
   members of their group. Marked stereotypy, self-
   stimulation, withdrawal, motor clumsiness, and
   abnormal vocal patterns were observed in the
   experimental monkeys. Parallel studies of EEG,
   evoked potentials, H reflexes, and blood serotonin
   were carried out (the last in collaboration with Mary
   Coleman of Children's Hospital, Washington, D.C.).
   Abnormal EEG tracings during slow-wave sleep of the
   isolates are consistent with a hypothesis of disturbed
   central regulatory mechanisms relating to arousal.
   Normal H-reflex recovery cycles are inconsistent with
   a hypothesis of cerebellar dysfunction. Serial
   determinations of platelet-bound serotonin revealed no
   significant differences between groups over the age
   range 12-18 mo.

615. Tarentino, S.J. Effects of cage confinement on social behavior in squirrel monkeys. Psychonomic Science, 1970, 20(5), 294-295.

   Data on the effects of cage confinement in squirrel
   monkeys were obtained as part of a larger study on
   social perception. Approach-withdrawal behavior
   was studied at three intervals over a 6-month period,
   during which the animals were confined in individual
   cages. At the sixth month a noticeable change in
   behavior occurred. During 20-min test intervals the
   animals displayed very strong attachment behavior.
   This was in marked contrast to their earlier behavior.

616. Testa, T.J., & Marck, D. The effects of social isolation on
sexual behavior in Macaca fascicularis. In S. Chevalier-Skolnikoff
and F.E. Poirier (Eds.), Primate bio-social development: Biological, social, and ecological determinants. New York: Garland
Publishing Inc., 1977, 407-438.

> The two studies presented in this paper are an attempt
> to analyze the sexual behavior of male-female dyads of
> crab-eating macaques (M. fascicularis) by first examining
> this behavior in ferally-reared male-female dyads, and
> then observing isolation-reared male-female dyads, and
> then observing isolation-reared animals in similar
> situations. In order to accomplish this, topographical,
> frequency, and sequential analyses of primate social
> interactions in both groups are performed.

617. Thompson, C.I. Time in test cage and behavior after amygdalectomy in infant rhesus macaque. Physiology and Behavior, 1969,
4, 1027-1029.

> Adolescent monkeys that had sustained bilateral amygdalec-
> tomy during infancy were observed immediately after
> transfer to a test cage, and again 24 hr after the move.
> Operated monkeys changed from one behavior to another more
> often than controls immediately after transfer to the test
> cage, but not after 24 hr. After 24 hr operated monkeys
> spent more time in oral and manual exploration of the cage
> than did the controls, but there were no significant group
> differences immediately after transfer. Operated monkeys
> thus were abnormal both in frequency of behavioral change
> and in the amount of time spent exploring, but both
> deficiencies would not have been detected had only one
> of the above test procedures been used. It is suggested
> that the amount of time spent in the test cage prior to
> testing may be an important variable in detecting subtle
> changes in behavior after brain damage.

618. Thompson, C.I., & Towfighi, J.T. Social behavior of juvenile
rhesus monkeys after amydalectomy in infancy. Psysiology and
Behavior, 1976, 17, 831-836.

> Six 3½-yr-old rhesus monkeys that had sustained a
> bilateral amygdalectomy in infancy were paired with
> 6 intact age-mates, and behaviors were observed immediately
> and after a 24-hr interval. Amygdaloid animals were

submissive to controls, and were hyperactive during
all tests, changing rapidly from one behavior to
another. Earlier studies with these infant-operated
monkeys showed that an abnormal fear of intact age-mates
first appeared during the second 6 mo of life. The
amygdaloid animals had been hyperactive when by themselves
as early as 6 mo of age, but there was no evidence of
hyperactivity in social situations during earlier tests.
The present paper presents the histology for these infant-
operated monkeys, and suggests that behavioral deficits
resulting from amygdalectomy in infancy increase in
severity with advancing maturation. The aberrations
observed by 3½ yr are similar to those observed after
amygdalectomy in mature animals.

619. Thompson, C.I., Bergland, R.M., & Towfighi, J.T. Social
and nonsocial behaviors of adult rhesus monkeys after amygdalectomy
in infancy or adulthood. Journal of Comparative and Physiological
Psychology, 1977, 91(3), 533-548.

At 6 yr of age six female rhesus monkeys that had sustained
bilateral amygdalectomy in infancy, and five intact controls,
were transferred to an observation cage where behaviors were
recorded while the monkeys were (a) alone, (b) paired
with unfamiliar stimulus animals, and (c) paired with
familiar monkeys from the opposite experimental group.
The five adult controls then underwent amygdalectomy, and
all tests were repeated with the infant- and adult-
operated animals. Infant-operated monkeys changed behaviors
more rapidly than did intact controls in social and nonsocial situations, and their activity levels were less
modified after a 24-hr period in the observation cage.
They were subordinate to intact controls but expressed
less fear than did controls when briefly placed with an
unfamiliar aggressive animal. Adult amygdalectomy produced
many changes in behavior, but these aberrations were
identical to those observed in like-age monkeys that had
been amygdalectomized in infancy. Infant-operated monkeys
demonstrated more behavioral deficits at 6 yr than they
had earlier in life.

620. Thompson, C.I., Schwartzbaum, J.S., & Harlow, H.F. Development of social fear after amygdalectomy in infant rhesus macaques.
Physiology and Behavior, 1969, 4, 249-254.

Infant rhesus monkeys were subjected to a bilateral
amydalectomy and were compared in group and individual
situations with sham-operated controls. During the
year following surgery significant group differences
developed in the expression of fear responses. In
general, the operated monkeys appeared less disturbed by
novel stimuli than the controls. In group situations,
however, the operated monkeys made more fear responses
than the controls, especially: (1) when the test groups
included unoperated monkeys; (2) when the length of the
test session increased; and (3) when the monkeys grew
older. Although other investigators have reported
virtually normal behavior following bilateral amygdalectomy
during infancy, it is clear from the present study that
some behavioral changes still occur. The results also
suggest that common generalization depicting amygdaloid
monkeys as "fearless" does not accurately describe their
long-term interactions with normal peers.

621. Thompson, C.I., Harlow, H.F., Blomquist, A.J., & Schiltz, K.A.
Recovery of function following prefrontal lobe damage in rhesus monkeys. Brain Research, 197, 35, 37-48.

Previous investigations have clearly demonstrated
that destruction of frontal granular cortex in
monkeys within the first 5 months of life is far less
detrimental to delayed response ability than is
damage sustained at a later time. Once this sparing
effect ceases to occur, however, performance after
surgery has tended to improve with age. In the present
investigation, monkeys operated at 5, 12, or from 18
to 24 months were compared with unoperated controls
on delayed response, object discrimination learning,
and ability to develop object discrimination or oddity
learning sets. Results showed that in all cases where
monkeys of different ages were unequally disrupted,
those tested soon after surgery at 12 months were
significantly inferior to those tested after surgery
at 18 or 24 months. These findings suggest that abilities
normally mediated by the frontal lobes can also be
mediated with somewhat less efficiency by other areas as
the remaining parts of the nervous system develop.

622. Tinklepaugh, O.L.   The self-mutilation of a male macaque rhesus monkey. Journal of Mammology, 1928, 9, 293-300.

This article presents a diary account of the self-mutilative behavior of a male rhesus monkey (Macaca mulatta) resulting from a stressful encounter with a female conspecific following monogamous attachment to another female (of a different species).

623. Tinklepaugh, O.L. & Hartman, C.G.   Behavior and maternal care of the newborn monkey (Macaca mulatta). Journal of Genetic Psychology, 1932, 40, 257-286.

This paper discusses the birth and development of rhesus monkeys (Macaca mulatta) under laboratory conditions. The infant-mother and maternal affectional systems are discussed.

624. Tucker, T., & Kling, A.   Effects of early and late frontal cortex lesions on delayed response in the monkey. Federation Proceedings, 1965, 24(2), 522.

Evidence exists for sparing of somato-sensory function after cortical lesions in the infant animal. Bilateral lesions of frontal granular cortex in the mature monkey result in an impairment in the capacity to perform delayed response. In an effort to examine the sparing for this function, bilateral ablations of frontal granular cortex were produced in three monkeys prior to the 4th postnatal day. Control lesions involving bilateral damage to the temporal lobes were produced in two monkeys at comparable ages. Three mature animals with bi-frontal injury comprised the adult-operated group. Testing procedures involved the initial learning of a color discrimination followed by a series of delayed response tests, with delay intervals ranging from zero to 40 sec. No marked differences were found among the experimental groups in learning the color problem. However, delayed response testing revealed a marked deficit among the adult-operated frontal animals relative to the performance of infant frontal-operates. These results suggest that delayed response capacity may be spared with frontal lesions sustained in infancy.

625.   Tucker, T.J. & Kling, A.   Differential effects of early and
late lesions of frontal granular cortex in the monkeys.   Brain
Research, 1967, 5, 377-389.

     Bilateral ablations of dorsolateral frontal granular
     cortex were performed in four monkeys prior to the 35th
     postnatal day and in three monkeys at 3 years of age.
     Subsequent testing procedures involved the learning of
     a color discrimination, a series of delayed response
     tests and a delayed alternation task. No differences
     were found between early- and late-frontals in learning
     the color problem. All animals were equally impaired in
     delayed alternation performance. Delayed response
     testing revealed a marked superiority for early-frontals,
     relative to the performances of late-frontals. Differential
     delayed response performances of the two frontal groups
     were not due to differences in either surface lesion or
     retrograde thalamic degeneration. It was concluded that
     either a compensatory neural reorganization or a progressive
     cerebral maturation might account for the sparing of
     delayed response capacity in the early frontal preparations.

626.   Turner, C.H., Davenport, R.K. Jr., & Rogers, C.M.   The effect
of early deprivation on the social behavior of adolescent chimpanzees.   American Journal of Psychiatry, 1969, 125, 1531-1536.

     Chimpanzees reared during early life in environments with
     social and perceptual restrictions are strikingly
     different from animals reared by their mothers in a natural
     habitat. As adolescents they avoid social contact and
     display little species-typical behavior; they play and
     copulate infrequently and do not groom. The authors
     found these abberations to be very resistant to modification
     by a variety of maneuvers, including contact with normal
     social partners, drugs, and experimental manipulations,
     and they discuss the implications of their lack of
     "therapeutic" success.

627.   Vessey, S.H.   Free-ranging rhesus monkeys: Behavioral effects
of removal, separation, and reintroduction of group members.
Behavior, 1971, 40, 216-228.

     Rhesus monkeys (26) varying in age, sex and social rank
     were removed from their free-ranging groups and held
     captive for periods of 1 to 103 days. The absence of the
     alpha male did not affect the group's home range on the

island habitats or its rank in the intergroup dominance
hierarchy. When released the monkeys interacted most with
those closest to them in rank, attacking those lower and
grooming those higher. Of 17 males, 8 failed to rejoin
their groups and became solitary or low-ranking in other
groups. Only 1 of 9 females failed to rejoin their group.
Males held captive for more than four weeks seldom rejoined
after release. Upon reintroduction alpha females experienced
more difficulty than lower-ranking females in maintaining
rank. This procedure of removing and reintroducing
members of free-ranging groups is of value in studying
social roles.

628. Waisman, H.A., & Harlow, H.F. Experimental phenylketonuria
in infant monkeys. Science, 1965, 147, 685-695.

Experimental phenylketonuria can be produced in infant
monkeys by feeding excessive quantities of L-phenylalanine
soon after birth. Both the phenylketonuric monkey and the
phenylketonuric human patient have elevated plasma levels
of phenylalanine, and monkey and human excrete almost the
same phenylalanine metabolites in the urine. Grand-mal
convulsions, observed in some children with phenylketonuria,
were also observed in the experimental animals. The
biochemical evidence was supported by the learning data.
The observed slowness in adapting to testing procedures, or
even failure to adapt, and the inadequate performance
suggest an intellectual deficit.

629. Wako, H., Hatakeyama, T., Kamihara, M., Wada, S., Honjo, S.,
Fujiwara, T. & Cho, F. Artificial nursing of new-born cynomolgus
monkeys as a model of the human infant and development of abnormal
behavior. Experimental Animal, 1975, 24(4), 161-171.

New-born cynomolgus monkeys were successfully reared by
artificial nursing that was started just after birth with
a 12% solution of a commercially prepared powdered-milk
(Yukijirushi, P 7a) containing 13.3g of protein per 100g.
Marked growth-retardation was observed in baby cynomolgus
monkeys fed on a 12% solution of the modified P 7a milk
containing only 6.6g of protein per 100g to which lactose
was supplemented to give a baby monkey the same caloric
value as that of the original P 7a milk. These artifically
reared cynomolgus monkeys manifested various kinds of
abnormal behavior such as self-sclasping, autism-like

self mouthing, huddling, sterotype rocking, head-
knocking, autoerotism, fear, aggression, etc..
Generally, development of these abnormal behaviors
was more noticeable in the monkeys nursed with a
milk bottle fixed to the side of a cage without human
contact than in the monkeys nursed by a care-taker with
bodily touching. These qualitative observational
results indicate that the new-born cynomolgus monkey
can be used as a model of the human baby for research
into the relationship between malnutrition and
abnormal physical and mental growth.

630. Waldrop, M.K. Responses of human-reared monkeys to strange
simian calls. Developmental Psychobiology, 1975, 8(3), 269-273.

The responses of 16 young (10-12 month and 19-20
month) hand-reared rhesus monkeys to rhesus vocalization
heard for the first time were recorded. The vocal
stimuli were: (1) a "clear call" of a noncolony adult;
(2) a "bark" of a noncolony adult; (3) a "bark" of a
noncolony juvenile; and (4) a "bark" of a colony juvenile.
Behavior was video-tape recorded for 5-min periods
preceding and following the calls and scored according to
standard categories. The characteristic response to the
stimuli was a sharp reduction in overt behavior. Reaction
to the adult sounds was greater than to the juvenile
sounds.

631. Wendt, R.H., Lindsley, D.F., Adey, W.B., & Fox, S.S.
Self-maintained visual stimulation in monkeys after long-term
visual deprivation. Science, 1963, 139, 336-338.

Newborn monkeys reared in darkness for 16 months, except
for daily 1-hour periods of exposure to unpatterned
light, were allowed to press a lever to obtain unpatterned light. The animals showed apparently
insatiable responding, at rates that were extremely
high as compared with rates for normally reared
control animals.

632. White, L.E., & Hinde, R.A. Some factors affecting mother-
infant relations in rhesus monkeys. Animal Behavior, 1975, 23,
527-542.

The effects of six independent variables on rhesus mother-
infant relations are examined by comparing sub-groups
of mother-infant pairs differing in only one of the
variables. Variables compared and discussed include
mother's parity, previous experience and social status,
sex of the infant and the presence of peers. All
variables except maternal parity appeared to affect the
relationship, but none produced effects that were wholly
consistent over all comparisons. Further tentative
analyses were made of some age effects and interactions
between variables which may have accounted for the
inconsistencies.

633. Willott, J.F., & McDaniel, J. Changes in the behavior of
laboratory-reared rhesus monkeys following the threat of
separation. Primates, 1974, 15(4), 321-326.

Four heterosexual pairs of three-year-old monkeys
were either repeatedly separated from each other
for 30-min or not separated. Prior to each
separation, a transfer cage was displayed at the front
of the cage to serve as a cue that the pair would
be separated. After only a few trials, the animals
displayed disturbance, particularly in the form of
stereotyped pacing prior to the separation. Practical
considerations related to laboratory methodology,
a theoretical discussion concerning the fear of
separation in monkeys and man, and the need for
attention to individual differences in response to
separation were emphasized.

634. Wilson, P.D., & Riesen, A.H. Visual development in rhesus
monkeys neonatally deprived of patterned light. Journal of
Comparative and Physiological Psychology, 1966, 61(1), 87-95.

Twelve rhesus monkeys, deprived of pattern vision from
the day of birth to 20 or 60 days of age, were tested
daily for various untrained visual responses and were
trained on form, striation, and brightness descriminations.
The 20- and 60-day-old visually deprived Ss were similar
to newborn monkeys in rate of learning visual discriminations
and in untrained visual behaviors. Acuity, brightness
difference thresholds, and generalization of learned
discriminations changed as Ss gained experience. Ocular
tracing of patterns, visual placing, and visual cliff
avoidance developed only after hours or days in patterned

light.   Savings in hours required relative to the normal
developmental time course gave evidence for maturational
as well as experiential contributions.

635. Windle, W.F. Asphyxia at birth, a major factor in mental
retardation: Suggestions for prevention based on experiments in
monkeys. In J. Zubin and G.A. Jervis (Eds.), Psychopathology of
mental development. New York: Grune and Stratton, 1967, 140-
147.

The problem of mental retardation is discussed with
regard to oxygen starvation and subsequent brain damage
seen in both human and nonhuman primates. The physiological
repercussions of asphyxia neonatorum are investigated in
various primate species including rhesus macaques
(Macaca mulatta). Several prophylactic techniques are
discussed.

636. Winter, P., Handley, P., Ploog, D., & Schott, D. Ontogeny
of squirrel monkey calls under normal conditions and under acoustic
isolation. Behavior, 1973, 47, 230-239.

Five mothers of squirrel monkey infants isolated from
other species members were muted by severence of their
vocal cords during pregnancy. After delivery, mother
infant pairs were brought up in an environment free of
any species-specific auditory input. One of these
infants underwent a deafening operation five days after
birth. In addition, two infants grew up under normal
conditions, i.e., exposed to species-specific
vocalization. Supplemental data were acquired from six
other infants, four of them normally raised and two
handraised. Sound spectrograms were taken over a period
up to six months in the case of the isolates and up to
17 months for the normal animals. Samples of this spectro-
graphical material were analyzed with respect to the form
of calls and to quantitative criteria, such as duration,
starting frequency, mid-frequency, and end frequency of
peep and cackle calls. Clear evidence is presented that
the vocal repertoire of squirrel monkey infants raised
under normal conditions and those raised in the absence
of species-specific auditory input are virtually
identical. Furthermore, comparison of the infants'
vocalization with those adult animals shows no
significant differences.

637. Wise, L.A. The effects of social experience, protein depri-
vation, and social coalitions on social dominance of rhesus monkeys.
(Doctoral dissertation, University of Montana, 1973) Dissertation
Abstracts International, 1974, 34(7).

    Three major experimental studies were conducted,
    all related to the basic topic of social dominance
    seen through food incentive competition in rhesus
    monkeys (Macaca mulatta). The first study involved
    measuring and comparing the stability and linearity of
    the dominance heierarchies within differentially reared
    groups, the second was an attempt to manipulate the
    competition outcomes between a group of low-protein-
    reared rhesus and a group of high-protein-reared monkeys,
    and the third looked at the effect of the presence of a
    stimulus-coalition subject upon the food competition of
    two other subjects for two of the groups. In general,
    all three studies yielded meaningful information to
    the study of social dominance in rhesus monkeys.

638. Wise, L.A., & Zimmermann, R.R. The effects of protein
deprivation on dominance measured by shock avoidance competition
and food competition. Behavioral Biology, 1973, 9, 317-329.

    Dominance behavior of rhesus macaques on either low or
    high protein diets was measured using three competition
    techniques. The appetitive measures consisted of food
    competition in the WGTA and food competition in a
    parallel box apparatus. The nonappetitive measure
    consisted of shock avoidance competition. The
    results indicated that rhesus monkeys raised on low
    protein diets were rated more dominant than rhesus
    monkeys raised on high protein diets on food
    competition and that high protein rhesus monkeys
    were more dominant on avoidance competition. All
    three measurement techniques were found to be stable,
    and the two appetitive measures were correlated
    highly.

639. Wise, L.A., & Zimmermann, R.R. Shock thresholds of low- and
high-protein reared rhesus monkeys. Perceptual and Motor Skills,
1973, 36, 674.

    Rhesus monkeys were separated from their mothers at
    90 days of age and maintained on a purified diet that
    was isocaloric and either contained 3% protein or

25% protein by weight. The diet was started at 120
days of age for eight Ss (4 high- and 4 low-protein Ss
age-paired) and at at 210 days of age for 10 Ss (4
high- and 6 low-protein Ss age-paired). Experimental
histories of all Ss (at the time of the measurements
listed below) indicated that all Ss had had previous
experience with electrical shock and all to approximately
the same extent. For each S the intensity of electrical
shock was varied by decreasing and increasing the intensity
of the shock (ranging from .01 to 2.0 ma). This was
repeated over a period of 14 days with one session a day
lasting about ½ hr. A Grason-Stadler shock generator
produced a scrambled shock through the floor, constructed
of long parallel metal bars. The test apparatus
consisted of a grey shuttle box, divided in half by a
barrier of metal rods (thus using only one side of the
shuttle box). The side of the shuttle box used was 63.5 cm
deep, 45.72 cm wide and 55.24 cm long, with a guillotine
door centered at one end which measured 27.94 cm up from
the metal rod floor and 22.86 cm wide. Observations of a
first noticeable response (visable to E, generally a foot
retraction or a digit extension) to shock intensity
yielded significant differences between both age-paired
groups of low- and high-protein-reared Ss. The 120-day
age-paired groups responded to a significantly lower
intensity of shock for the low-protein-reared Ss than
high-protein-reared Ss (t = 2.3, df = 6, p < .05). The
first noticeable response to shock ranged from .10 to
.13 ma (M = .12) for the 120-day low-protein group and
from .13 to .25 ma (M = .19) for the 120-day high-protein
group. The 210-day age-paired group also showed a significantly lower intensity of shock response for the low-protein-
reared Ss than the high-protein-reared Ss (t = 2.4, df = 8,
p < .05). The first noticeable response to shock
ranged from .06 to .25 ma (M = .17) for the 210-day
low-protein group and .20 to .30 ma (M = .26) for the 210-
day high-protein group. Variables accountable for these
discrepancies in threshold for shock-intensity could be in
differences in gross body weight, variations of thickness of
cutaneous and dermal layers, deviations in neuron
excitability and others. Physiological and behavioral
measurements are currently being studied .

640. Wolfheim, J.H., Jensen, G.D., Bobbitt, R.A. Effects of group
environment on the mother-infant relationship in pigtail monkeys
(Macaca nemestrina). Primates, 1970, 11(2), 119-124.

Behavior of six <u>Macaca nemestrina</u> mother-infant pairs was
observed in two different environments: three pairs were
part of a large group in a compound and three pairs
were housed in individual laboratory cages. At weeks 14-15
of the infants' lives, group mothers were more retentive
of their infants than caged mothers; group infants spent
more time in ventral contact with mother and nursed more
than did caged infants. The authors conclude that the
greater dependence of an infant raised in a potentially
dangerous group setting is due to a more protective
mother rather than to a more fearful infant.

641. Yarrow, L.J. The crucial nature of early experience. In
D.C. Glass (Ed.), <u>Biology and behavior-environmental influences</u>.
New York: Rockefeller University Press and R. Sage Foundation,
1968, 101-113.

This paper presents a very broad examination of early
experience and its relationship to the developing
organism. Various designs and measurement techniques
are discussed.

642. Yellin, A.M. Circadian periodicity in the rhesus monkey
(<u>Macaca mulatta</u>) under several conditions, both in isolation and
in a group situation (Doctoral dissertation, University of
Delaware, 1970). <u>Dissertation Abstracts International</u>, 1971,
<u>3</u>(9-B), 71-6477, 5685B.

The activity of rhesus monkeys was measured under
several environmental conditions to determine their effect
on the monkeys' circadian periodicity. The monkeys were
isolated in separate rooms and placed on a schedule of 12
hours of light followed by 12 hours of darkness (LD=12:12)
for 30 days. The activity of the monkeys under this
condition exhibited great photoperiodic dependency, with
activity onset and termination linked to the onset and
termination of the light period. The shape of the daily
activity curve was consistent throughout the LD=12:12
condition and across subjects. The isolated monkeys were
next exposed to two constant environments which were
controlled for all known synchronizing signals. Under
conditions of constant light (LL =40 days), the circadian
frequencies exhibited by the subjects deviated from 24
hours. Marked individual differences were observed in
the speed with which the shape of the daily activity curve
became modified from the one exhibited under the LD=12:12

schedule. The individual monkeys were then permitted
to self-regulate their own exposure to light and dark
(SLD-I$_1$ = 40 days). Under this condition marked changes
were apparent in the activity/rest cycle, with the
isolated monkeys spending most of the time in darkness.
The circadian frequency became modified from what it had
been under the previous (LL) condition. There were
indications that under this condition the circadian
frequency may become highly irregular, or disappear
altogether. The subjects were next placed together
in a colony room (SLD-Gr = 40 days). Each monkey was
permitted to regulate the lights in the room independently
of the other monkeys. Under this condition mutual
entrainment was observed. The circadian frequencies
of all the subjects became similar. In 90% of the days
it was one subject who first turned the lights on for
the main active period. This subject was determined to
be the most aggressive of the monkeys. The lights were
kept on much longer in the group condition, relative to the
SLD-I$_1$ condition. At the termination of the group condition
the monkeys were again placed in isolation and permitted
self-regulated exposure to light and dark (SLD-I$_2$ = 30
days). This condition provided additional evidence for
the entrainment apparent under the group condition. All
subjects began their "free-running" from the same point
in time after being placed back in isolation. This
condition also provided further indications of the
lingering effects of a previous condition on a current
one. The circadian frequency of all the subjects was
different from what they had exhibited during the first
self-selection condition (SLD-I$_1$), and close to the
circadian frequency observed under the group condition.
The group condition and the isolation condition following
it demonstrated sociality to be a powerful synchronizer of
the biological clock of primates.

643. Yellin, A.M., & Hauty, G.T. Activity cycles of the monkey
(Macaca mulatta) under several experimental conditions, both in
isolation and in a group situation. Journal of Interdisciplinary
Cycle Research, 1971, 2(4), 475-490.

Rhesus monkeys (Macaca mulatta) were subjected to several
successively applied, environmental conditions. Their
activity was monitored by means of ultra-sonic movement
detectors, and the data were submitted to a least squares
spectral analysis. The purpose of this series of experiments was to monitor and compare the activity cycles of

isolated unrestrained monkeys, and to assay the role of sociality as an entrainer of general activity in the absence of other known synchronizing signals. The data supported the notion of social entrainment in group situations.

644.   Young, L.D., Suomi, S.J., Harlow, H.F., & McKinney, W.T., Jr.   Early stress and later response to separation in rhesus monkeys.   American Journal of Psychiatry, 1973, 130(4), 400-405.

The hypothesis that separation in early life predisposes one to later psychopathology was tested using two groups of rhesus monkeys of similar age.   The experimental group had undergone the stress of early life separation and confinement in a vertical chamber.   The experimental animals responded to separation and reunion with increased self-mouthing, self-clasping, huddling, and rocking.   The control group responded to separation with increased locomotion and responded to reunion with increased contact clinging and proximity behaviors. The implications of these findings are discussed.

645.   Young, L., McKinney, W.T., Jr., Lewis, J.K., Breese, G.R., Smith, R.D., Mueller, R.A., Howard, J.L., Prange, A.J., & Lipton, M.A.   Induction of adrenal catecholamine synthesizing enzymes following mother-infant separation.   Nature New Biology, 1973, 246, 94-96.

Studies were undertaken to determine whether the 'protest' stage following maternal separation in infant rhesus monkeys (Macaca mulatta) was accompanied by alterations in levels of catecholamine synthesising enzymes as well as choline acetylase activity in the adrenals and the superior cervical ganglion.   To evaluate the central effects of this procedure, levels of brain biogenic amines and tyrosine hydroxylase activity were also measured in the same infants.   The data indicate that mother-infant separation stimulates adrenal catecholamine synthesis in the infant rhesus, and that this effect is neurally mediated.   The findings of increased tyrosine hydroxylase and choline acetylase levels in the cervical ganglia also indicate increased sympathetic activity in response to this procedure.

646.   Zimmerman, R.R.   Form generalization in the infant monkey. Journal of Comparative and Physiological Psychology, 1962, 55(6), 918-923.

    Seventeen rhesus monkeys were divided into four groups.
    Two groups were trained on a triangle-circle
    discrimination problem to a criterion of 21 correct out
    of 25 responses on 2 successive days and then tested
    for primary stimulus generalization. A third group
    of subjects was trained to this same criterion and then
    received additional training on 10 pairs of stimuli that
    were similar but not identical to the generalization
    stimuli. A fourth group of subjects received 200
    overtraining trials on the original pair of stimuli.
    The results are discussed with regard to the infant
    monkeys' abilities to generalize.

647.   Zimmerman, R.R.   Early weaning and weight gain in infant rhesus monkeys.   Laboratory Animal Care, 1969, 19, 644-647.

    Infant rhesus monkeys were separated from their
    mothers shortly after birth, housed in individual
    cages, and maintained on a Similac formula. At 30
    days of age, solid food in the form of a purified
    high protein diet was introduced. Starting at 70 days
    of age, one group of 6 monkeys was fed a solid food
    diet only, while a second group of 12 monkeys continued
    on the liquid diet and solid food. At 120 days of age,
    a second group of 6 infants was now fed only the solid
    food, while a third group of 6 continued on the liquid
    and solid diet. The weaning procedures at 70 days of
    age produced a significant reduction in rate of weight
    gain. No significant changes were found at 120 days of
    age.

648.   Zimmerman, R.R.   Effects of age, experience, and malnourish-ment on object retention in learning set.   Perceptual and Motor Skills, 1969, 28, 867-876.

    Repeated presentation of 100 6-trial problems produced
    significant retention of individual object discriminations
    in both 25-mo.-old and 6-mo.-old infant monkeys. The
    amount retained was a function of amount learned in
    any one sequence and was independent of age and
    experience. Deprivation in the form of malnutrition
    of 6-mo.-old monkeys produced superior performance

in intraproblem learning and subsequent retention, but
the retention loss remained constant. A 3-mo.
retention interval produced a reduction in retention
of individual items in the 25-mo.-old S̲s̲ but not in
the younger animals. Return to a normal diet resulted
in a decrease in performance, a return to performance
normal for 6-mo.-olds after a 3-mo. interval. It
appears that retention loss and the superior performance
of the deprived group can be traced to changes in
stimulus-preference error. The results are consistent
with experiments using short memory intervals.

649. Zimmerman, R.R. Transfer of a learned form discrimination
in the infant and adult monkey. Developmental Psychobiology,
1970, 3(1), 35-41.

Eight infant monkeys with a mean age of 67 days,
range 50 to 84, and 8 adult monkeys were trained to
discriminate an equilateral triangle from a circle.
After reaching the criterion of learning, they were
subjected to a series of transfer tests containing
20 different stimulus pairs. The groups did not
differ significantly in rate of learning or transfer
scores. Adults showed significant interproblem
transfer to stimuli that had high generalization
scores in other studies. Significant transfer was
found to outline figures, intensity reversal and
no consistent effect in responses to the sides of the
triangle as opposed to the angles appeared.

650. Zimmerman, R.R. Abnormal social development of protein-
malnourished rhesus monkeys. Journal of Abnormal Psychology, 1972,
80(2), 125-131.

Monkeys fed diets deficient in protein (3.5% Casein)
exhibited social behaviors qualitatively and
quantitatively different from monkeys fed diets
sufficient in protein (25% Casein). In general, the
protein-malnourished animals showed a lack of reciprocal
responsiveness to social initiations, aggression, and
a predominance of nonsocially directed behavior when
compared to the controls. The results are discussed in
terms of a nutritional-environmental interaction that
produces abnormal and possibly long-term deficits in
social development.

651.   Zimmerman, R.R.   Reversal learning in the developing malnourished rhesus monkey.   Behavioral Biology, 1973, 8, 381-390.

    Two experiments were run in which monkeys subjected to a
    low protein diet were tested on a series of object-
    discrimination learning and reversal problems.   In the
    first experiment year old monkeys were tested before going
    on the diet.   At 14 months a diet containing 3% protein
    was introduced and one month later testing was resumed.
    No significant changes in performance were found.   In the
    second experiment 150 day-old monkeys that had been on the
    low protein diet since they were 90 days old were
    compared with controls reared on normal diets on the
    learning and reversal problems.   The low protein animals
    were inferior on learning the reversal problems.

652.   Zimmermann, R.R. & Strobel, D.A.   Effects of protein malnutrition on visual curiosity, manipulation, and social behavior
in the infant rhesus monkey.   Proceedings of the 77th Annual
Convention of the American Psychological Society, 1969, 241-242.

    Eight infant rhesus monkeys born under laboratory conditions
    and separated from their mothers 110 to 180 days after
    birth were housed in individual wire cages and given
    access to standard monkey chow.   One group of four monkeys
    (the low protein group) was approximately one year old
    at the beginning of the experiment, while a second group
    of 4 monkeys (the high protein group) was 7 months of age
    on the average.   All groups were fed high protein diets
    and tested in a visual exploration apparatus.   Approximately
    2 months after the introduction of the high protein diet,
    4 monkeys, 14 months of age, were placed on a low protein
    diet.   All subjects were weighed daily, blood samples
    were taken, and total serum protein levels were determined.
    Visual curiosity, manipulation, and social behavior were
    observed during the deprivation period.   The results
    demonstrated that the low protein group was not significantly
    different from the high protein group on visual curiosity,
    but differed markedly on manipulation and social behaviors
    from the high protein group.   These differences were
    manifested by lower activity levels and minimal social
    contact.

653. Zimmerman, R.R., Geist, C.R., & Ackles, P.K. Changes in
the social behavior of rhesus monkeys during rehabilitation from
prolonged protein-calorie malnutrition. Behavioral Biology, 1975,
14, 325-333.

    Social behavior of rhesus macques placed on repletion diets
    after periods of up to 3 years of chronic protein mal-
    nutrition was measured in a playroom situation. Observa-
    tions of experimental and control groups were made 5 days
    a week for 5 weeks prior to and beginning again 30 days
    after the situation of repletion. The behaviors observed
    prior to nutritional rehabilitation were similar to that
    reported in previous experiments. Nutritionally deprived
    animals showed higher rates of aggression, less play
    behavior, less tactual contact and a greater amount of
    nonsocial behavior. The animals undergoing nutritional
    rehabilitation showed significant increases in frequency
    and duration of approach-play, significant reductions
    in nonsocial behaviors, and rates of aggression. Although
    the behavior patterns still did not reach control levels,
    the changes in the previously malnourished monkeys
    apparently made these animals more adaptive in the social
    situations.

654. Zimmermann, R.R., Geist, C.R., & Strobel, D.A. Behavioral
deficiencies in protein-deprived monkeys. In G. Serban (Ed.),
Advances in behavioral biology, vol. 14, Nutrition and mental
functions. New York: Plenum Press, 1975, 33-64.

    This paper discusses a series of experiments concerned
    with the effects of protein-calorie malnutrition on
    the behavioral development of the rhesus monkey
    (Macaca mulatta). The function of an animal model
    as an analog to the human condition is emphasized.

655. Zimmermann, R.R., Geist, C.R., & Wise, L.A. Behavioral
development, environmental deprivation and malnutrition. In
G. Newton & A.H. Riesen (Eds.), Advances in psychobiology (Vol. II).
New York: Wiley, 1974, 133-193.

    This paper presents a very extensive analysis of the
    effects of nutrition on behavioral development under
    various levels of social stimulation and paucity. Sixty-
    two laboratory-born rhesus monkeys (Macaca mulatta) were
    placed in experimental conditions designed to evaluate

the relationship between behavioral development, protein-
calorie malnutrition, and social rearing conditions. Curiosity, manipulation, responses to novel stimuli, learning
ability, attention, and social behavior are thoroughly
discussed with regard to the aforementioned independent
variables.

656. Zimmermann, R.R., Strobel, D.A., & Maguire, D. Neophobic
reactions in protein malnourished infant monkeys. Proceedings of
the 78th Annual Convention of the American Psychological Association,
1970, 197-198.

Ten infant rhesus monkeys (Macaca mulatta) were separated
from their mothers 110-180 days after birth and housed in
individual wire cages with free access to a purified
22% protein diet. At 210 days of age, 6 of the subjects
were placed on a purified diet containing 3% protein but
equal in calories to the 22% diet. All subjects were then
adapted to the Wisconsin General Test Apparatus and
presented 5 six-trial problems per day, 5 days a week.
Subjects were tested at 3 months and 11 months following
diet initiation. During the 11th month testing period a
seventh-trial was presented with a new discriminative
stimulus. The results of this investigation are discussed
with regard to neophobic or neophilic reactions to novel
stimuli.

657. Zimmermann, R.R., Strobel, D.A., Steere, P., & Geist, C.R.
Behavior and malnutrition in the rhesus monkey. In L. Rosenblum
(Ed.), Primate Behavior, vol. 4 Academic Press, 1975, 242-306.

This paper describes research which evaluates behavior
changes during the period of protein deprivation and
reexamines these behaviors after rehabilitation. Inherent
in the philisophy of this research program is the notion
that as behavior is altered during the development period
of the animal, adult or maturing behavior will also be
altered. Results from these investigators indicate that
motivational and social variables are dramatically affected
by malnutrition. These are drastic and widespread
abnormalities that develop and persist over a long period
of time. It is possible that there are a series of mutually
interacting patterns of neophobia, food preoccupation,
and immaturity that contribute to the behavior abnormalities
found in the low-protein animal.

658.   Zimmermann, R.R., Strobel, D.A., Maguire, D., Steere, R.R.,
& Hom, H.L.   The effects of protein deficiency on activity, learning,
manipulative tasks, curiosity, and social behavior of monkeys.   In
J. Bruner, A. Jolly, and K. Sylva (Eds.), Play -- its role in
development and evolution.   New York: Basic Books, 1976, 496-511.
511.

This article discusses research on the effects of protein
malnutrition on laboratory-reared monkeys.   The primary
preliminary findings of the present paper were the drastic
and widespread abnormalities in social behaviours shown by
those monkeys with induced protein malnutrition.   Positive
social behaviours such as approach play are depressed in
the low-protein monkeys, while negative social behaviours
such as fear and aggression are accentuated in the malnourished monkeys.   It is possible that there are a series
of mutually interacting patterns of neophobia, food
preoccupation, and immaturity which make some contribution
to the behavioural abnormalities observed in the low-
protein monkeys.

PREFACE TO INDEXES

Although all topic indices are somewhat subjective, and as such
reflect the authors' personal choices with respect to what will or
will not be included, the subject matter of this bibliography presented a particular challenge. Because many of the references are
concerned with separation and isolation as variables which can
affect nonhuman primates and virtually all subjects experience some
effects of either one or both of these treatments, a decision was
made not to include, as separate topics, either mother-infant separation or partial isolation. Those studies which deal directly with
mother-infant behavior are indexed under mother-infant interactions,
and a separate category for total isolation (Isolation, Total) is
included.

You will note that in addition to the usual author and subject
indices, we have included a species index and a special index for
review articles. It is hoped the latter will enable an investigator
or student to familiarize themselves with a particular area in a
more expeditious fashion than that suffered in the usual literature
search.

AUTHOR INDEX

Aakre, B., 1

Ackles, P.K., 653

Adey, W.B., 631

Agar, M., 2

Ainsworth, M.D.S., 3

Akert, K., 4, 212

Albert, S., 5

Alexander, B.K., 6, 7, 8, 555

Allen, J.R., 9

Allyn, G., 10

Alpert, S., 500

Altman, S.A., 11

Anderson, C.D., 12, 13

Angermeier, W.F., 14-22

Arling, G.L., 24, 26, 221, 436, 517

Ausman, L.M., 27

Baker, R.P., 468

Baldwin, D.V., 28

Bauer, J., 29, 237

Baysinger, C.M., 30-32, 66, 217

Beasley, J.W., 33

Becker, J.D., 47

Beckett, P.G.S., 34

SPECIES INDEX

Assamese Macaque (<u>Macaca assamensis</u>), 133

Baboon (<u>Theropithecus, Papio</u>), 72, 331, 419, 509

Bonnet Macaque (<u>Macaca radiata</u>), 5, 143, 283-285, 288, 290, 293-295,
    450, 497, 499, 500, 503, 504, 506, 507

Cebus Monkey (<u>Cebus albrons, cebus apella</u>), 112-114

Chimpanzee (<u>Pantroglodytes</u>), 37, 48, 50, 91, 94, 98-104, 129, 138,
    139, 141, 162, 233, 234, 240, 261, 298, 316, 318, 343, 346, 354,
    367, 368, 371, 372, 404-407, 409-413, 448, 453, 470, 480, 481,
    482-486, 489-491, 543, 626

Crab-eating Macaque (<u>Macaca fascicularis/irus</u>), 38, 39, 41-44, 46,
    47, 142, 282, 326, 467, 550, 616

Cross Species, 1, 2, 26, 69-70, 132, 134, 139, 157, 160, 171, 173,
    192, 331, 344, 348-352, 360, 363, 369, 427, 442, 457, 471, 483-
    485, 497, 552, 641

Gorilla (<u>Gorilla gorilla</u>), 447, 449

Inter-species, 45, 283-285, 288, 290, 293, 294, 308, 500, 503, 504,
    506, 507, 528, 529, 535-537, 540, 578

Japanese Macaque (<u>Macaca fuscata</u>), 297, 420

Marmoset (<u>Callithrix</u>), 49

Patas Monkey (<u>Erythrocebus patas</u>), 75, 468

Pig-tail Macaque (<u>Macaca nemestrina</u>), 59, 60, 75, 126, 133, 238,
    262, 263, 266-272, 283-285, 288, 290-294, 475-479, 497, 500, 503,
    504, 507, 528, 529, 535-537, 539-540, 560-562, 640

Rhesus Macaque (<u>Macaca mulatta</u>), 3, 6-9, 11-25, 27, 29-36, 40,
    51-58, 61-73, 66-68, 71, 73, 76-79, 81-86, 88-90, 92, 93, 95-97,
    105-111, 115, 116, 118-125, 127, 128, 130, 131, 135-137, 140,
    143, 144, 146-153, 155, 156, 158, 159, 161-170, 172, 173, 175-
    231, 235, 236, 239, 241-244, 246-258, 260, 275, 296, 300-304,
    312, 315, 319-326, 328, 329, 331-342, 345, 347, 348, 351, 353,
    355-366, 370, 375, 380-385, 388-395, 398-401, 403, 414-419, 422-
    427, 429-441, 446, 451, 452, 454-457, 459, 465, 469, 472-474,
    484, 485, 488, 492, 493, 496, 501, 502, 510-512, 517-534, 536-
    538, 540-542, 544-549, 551-558, 563-570, 572, 574, 575, 578-583,
    585, 586, 588-598, 600, 602-614, 617-623, 627, 628, 630-635, 637-
    639, 642-658

INDEX OF SURVEYS